SIMPLY EINSTEIN

SIMPLY EINSTEIN

Relativity Demystified

• • •

RICHARD WOLFSON

W • W • NORTON & COMPANY

NEW YORK • LONDON

For information about permission to reproduce selections from this book,
write to Permissions, W. W. Norton & Company, Inc., 500 Fifth Avenue,
New York, NY 10110

The text of this book is composed in Sabon with the display set in Futura
Regular and Light
Composition by Gina Webster
Manufacturing by Quebecor Fairfield
Book design by Margaret M. Wagner
Production manager: Julia Druskin

Library of Congress Cataloging-in-Publication Data

Wolfson, Richard.
   Simply Einstein : relativity demystified / Richard Wolfson.— 1st ed.
       p. cm.
Includes bibliographical references and index.
   ISBN 0-393-05154-4 (hardcover)
   1. Relativity (Physics)—Popular works. I Title: Relativity demystified.
   II. Title.
QC173.57 .W65 2003
530.11—dc21   2002002984

W. W. Norton & Company, Inc.
500 Fifth Avenue, New York, N.Y. 10110
www.wwnorton.com

W. W. Norton & Company Ltd.
Castle House, 75/76 Wells Street, London W1T 3QT

1 2 3 4 5 6 7 8 9 0

For Irving Wolfson
and Leonard Swift

CONTENTS

PREFACE

• • •

Have you ever heard it said of a difficult idea that "it would take an Einstein to understand this"? What could be more incomprehensible to us non-Einsteins than Albert Einstein's own work, the theory of relativity?

But relativity *is* comprehensible, and not just to scientists. At the heart of relativity is an extraordinarily simple idea—so simple that a single English sentence suffices to state it all. Some consequences of that statement are disturbing because they violate our deeply held, commonsense notions about the world. Yet those consequences flow inexorably from a single principle so simple and obvious that it will take me just a few pages to convince you of its truth.

This book's title, *Simply Einstein*, reflects the fact that the basic ideas of Einstein's relativity are accessible to nonscientists and make eminent sense. Even relativity's startling implications about the nature of space, time, and matter follow so directly from those basic ideas that they, too, become not only comprehensible but also logically inevitable.

Relativity is behind many of the hot topics at the frontiers of modern physics, astrophysics, and cosmology—topics ranging from black holes to the ultimate fate of the Universe to the prospects for time travel. I'll touch on these topics here, and you'll see how they flow from the essential ideas of relativity. But my main purpose is not to explore the latest frontiers of physics. There are plenty of good books on those topics, and I've included some in the Further

Readings. Rather, this is a book that aims to give you, its reader, a clear understanding of just what it was that Einstein said about the ultimate nature of physical reality. To help you get there, we'll be exploring together the history of ideas that culminated in Einstein's simple but remarkable vision. Then you'll see how that vision alters your commonsense notions of space and time in ways that would let you travel a thousand years into the future in just a few short hours. You'll come to a new understanding of "past" and "future" that might surprise historians, and you'll begin to feel at home in the four-dimensional universe of relativistic spacetime. Along the way I'll anticipate your frequent questions: Why can't anything go faster than light? Will I really age more slowly, or is this just something that happens to physicists' clocks? Can I go backward in time? What does $E = mc^2$ really mean? Finally, in the end, we'll return to some of those contemporary hot topics that show just how prescient was Einstein's visionary insight.

You don't need to do math to grasp the essence of Einstein's relativity, and you don't need math to understand this book. Occasional numbers can help make some points more concrete, and I'll use them sparingly. What's important here are the big ideas—and they're all expressed in words. Grasp those ideas, and you know what Einstein's relativity is all about. Enjoy!

SIMPLY EINSTEIN

● ● ●

CHAPTER 1

THE SELF-CREATING UNIVERSE AND OTHER ABSURDITIES

•   •   •

Could the Universe have created itself? What an absurd idea! Did the Universe even have a beginning? That question, too, has an absurd ring. If there was a beginning, what came before? Wasn't that part of the Universe too? Or has the Universe always existed, begging the question of its own origin?

Whatever the answers to these questions, modern astrophysics makes one thing clear: our Universe hasn't existed forever unchanged. Rather, it's evolved from an earlier state of extreme temperature and density. Some 14 billion years ago, all the "stuff" that makes up ourselves, our planet Earth, and all the stars and galaxies was crammed into a volume far smaller than a single hydrogen atom or even the tiny proton at its core. The expansion of that extreme state is the Big Bang that describes the Universe's subsequent evolution and ultimately accounts for the origin of stars, galaxies, planets, and intelligent life.

What came before the Big Bang? What created that early, extreme state? We're back to the primordial question: Did the Universe have a beginning, or has it always existed—albeit an existence marked by evolutionary change?

To some cosmologists—scientists who concern themselves with the origin and evolution of the Universe—the start of the Big Bang marks the start of time itself. For them, it makes no sense to ask what came before because the concept of "before" is meaningless if there's no such thing as time. Others have envisioned an ever existing Universe that undergoes a series of oscillations. Each begins with a Big Bang and subsequent expansion—the phase we're now in—then eventually contracts toward a Big Crunch of extreme density and temperature that starts another cycle.

In 1998 Princeton physicist J. Richard Gott and his student Li-Xin Li published a novel answer to the ultimate question of the Universe's origin. Their paper, "Can the Universe Create Itself?," shows how the laws of physics may allow a *time loop*, in which time goes round and round in a circlelike structure rather than advancing inexorably into a never-before-experienced future. Like Bill Pullman's character in the film *Groundhog Day*, an occupant of the time loop might go to bed at night and wake up on the morning of the day before! The "new" day would unfold, night would come, and again the morning would bring the already familiar day. In this loop, time advances circularly into a future that is a recycling through past events. There's no earliest event, any more than any point on a circle can be called the beginning of the circle or any point on Earth's surface is the place where our planet "starts."

Gott's time loop doesn't sound like a description of our Universe, but hold on—there's more. Time, in Gott's theory, can branch, providing different paths to different futures. Gott envisions a universe whose earliest epoch includes a time loop. Every point on the loop both precedes and follows every other point. There is no beginning instant, because one can always trace time further back round the loop. But there's a branch out of the time loop, a branch into a more normal realm of time that advances, without repetition, to a future of never-before events. That's the kind of time we know, with an as-yet-unknown, yet-to-occur future. Figure 1.1 depicts Gott's time-loop, multibranched universe.

It's the branching that reconciles Gott's time loop with the more ordinary time we experience today. The time loop unambiguously precedes our present, and in that sense it's closer to the beginning. But trace time backward, through the branch and onto the loop. You can keep tracing back but you'll never find a beginning. Instead, events repeat as your historical exploration circles backward around the loop. There's no one event that marks the creation

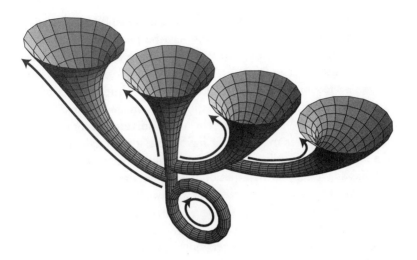

Fig. 1.1 Gott's time-loop universe. Arrows represent multiple directions of time, including the circular time loop at the beginning, in which time goes round and round in an ever repeating sequence. Circular cross sections of the "trumpets" represent position in a single spatial dimension, and each trumpet is its own universe. Is this absurd? Maybe not, says the theory of relativity. (Adapted with permission from J. Richard Gott and Li-Xin Li, "Can the Universe Create Itself?" *Physical Review D* 58 (1998), p. 3501.)

of this universe. Every event on the time loop precedes every other event, and in that sense Gott's universe creates itself. Absurd!

How Many Universes Did You Order?

Surely, "The Universe" encompasses all that there is. That's the root meaning of the word, as in "universal." But not according to Stanford University cosmologist Andrei Linde. For the Russian-born Linde, our Universe is but one small branch of a possibly infinite *Multiverse*. What we think of as the Big Bang origin and evolution of *the* Universe is, to Linde, simply the "budding" and subsequent expansion of a new branch from a pre-existing cosmos. That branch is our Universe. Other branches are different universes, each of which has had its own big bang and its own evolutionary scenario. Remarkably, each universe may even have its own laws of physics. The budding that produces a new universe may result in mutations from the laws that govern the parent branch. Together, all these interconnected universes form the Multiverse or, in Linde's more dynamic phrasing, the "self-replicating inflationary universe." Our own Universe may someday spawn new buds that become entire universes; in fact, it may already have done so. It might not even take much effort to initiate such a bud. Cosmologist Alan Guth of MIT has suggested that with an ounce of material, crushed to high enough density, you might start a new universe right in your own garage! Perhaps we and our whole Universe are just the results of someone's experimentation in another branch of the Multiverse.

Linde's Multiverse provides yet another answer to the question of the Universe's origin. *Our* Universe, according to Linde, clearly had a beginning in the budding event that was the start of *our* Big Bang. But that budding occurred from one branch of a Multiverse that may have existed forever—as if the structure in Figure 1.1, instead of starting with the time loop, just continued backward forever in a jumble of budding and branching universes. That self-replicating Multiverse in some ways resembles a biological system. It's forever spawning new buds—"baby universes"—some of which grow to become full-blown universes like our own, which then produce their own babies. Others are stillborn, withering to collapse before they've had a chance to evolve complex structure and intelligent life. Universes come and go, so there are multiple beginnings. Creation isn't a one-time story. But the Multiverse persists forever, and, despite the birth and demise of individual universes, the large-scale picture may remain unchanged for eternity.

Contact!

In the film *Contact*, based on Carl Sagan's novel with the same title, actress Jodi Foster plays the first astrophysicist to detect interstellar signals from an advanced civilization. The signals convey a message—instructions for building some sort of machine. Machines are built, against a backdrop of political and religious intrigue, and eventually Foster's character boards one for a ride into the unknown. The machine takes her through a wormhole in spacetime and deposits her on a distant world where she learns that entire galaxies are actually cosmic engineering projects. She rides the wormhole machine back to Earth only to find that no one believes her story because, for folks on Earth, no time has elapsed while she was ostensibly touring the cosmos.

Contact looks like science fiction. But many of its key ideas, including wormholes through space and time, are based in sound physics. In fact, author Sagan—a scientist himself—consulted colleagues about the validity of the sci-fi ideas at the basis of Contact. As a result, Sagan's novel spurred a flurry of interest in wormholes and in the possibility of time travel. By the turn of the century, leading researchers had published scores of papers on these subjects in the most respected physics journals. Some show how wormholes might connect seemingly distant parts of the Universe that are actually less than an inch apart in a hidden dimension. Others debate the mathematical possibility and philosophical implications of time machines that might let us travel into the past. That's getting pretty speculative, but another form of time travel is solidly established. Read on.

Escape to the Future

In the 1970s, scientists sent a very accurate atomic clock on a trip into the future. How? By flying the clock around Earth on commercial airplanes. When the clock returned to its starting point, it showed less elapsed time than a companion clock that hadn't made the journey. So what? That time difference means the traveling clock had somehow jumped into the future, arriving back at its starting place at a time that was further advanced than its own reading would suggest. For the traveling clock the difference amounted to some 300 billionths of a second. No big deal!

But we're convinced that the same idea would work in a more dramatic context, allowing you to "leapfrog" into the distant future.

Here's the scenario: You and I, who are about the same age, collab-orate in building a high-tech spaceship capable of traveling at close to the speed of light (about 186,000 miles per second). You board the ship, zoom off to a star in our galactic neighborhood, and return. As far as you're concerned, the trip takes a few days. But on return, you find me some 20 years older than when you left. You pick up a news-paper, and it's dated 20 years after your departure. You look around and see that planet Earth and all of human society have advanced 20 years during your several-days trip. Somehow you, like the atomic clock, have jumped into the future. This time the jump isn't a negli-gible fraction of a second but a goodly chunk of a human lifetime. Take that spaceship further—say, to the center of our Milky Way galaxy—and when you return, weeks later as far as you're con-cerned, you'll find yourself some 60,000 years in the future!

Alas, there's no going back—at least not with this form of time travel. If you don't like what you find thousands of years in the future, you can either put up with it or jump further into the future.

The Universe as Telescope

Much of what we know about the Universe beyond our home plan-et comes from telescopes—instruments that collect and analyze light and other forms of radiation from the cosmos. Before the early twentieth century, telescopes were too weak to see very clearly beyond our own Milky Way galaxy. That changed with the comple-tion of the 100-inch Mount Wilson telescope in 1917, and within 10 years observations from Mount Wilson had radically enlarged humankind's conception of the Universe—a conception that includ-ed, for the first time, hints that the Universe had a beginning.

Mount Wilson reigned for three decades as the world's most powerful telescope, until the 1949 opening of the 200-inch Mount Palomar instrument quadrupled astronomers' powers of observa-tion. (Those 100-inch and 200-inch figures are the diameters of the telescopes' mirrors; the amount of light they collect depends on the mirror area, which quadruples for each doubling of its diameter.) Today's largest ground-based telescopes have mirror diameters of 10 meters, or nearly 400 inches. With their huge but flexible mirrors, prime mountaintop sites, and a host of advanced technological features, these modern instruments greatly extend our vision of the Universe.

The 1990 launch of the Hubble Space Telescope gave astronomers an exciting new tool for probing the Universe. Its location above the distorting effects of Earth's atmosphere more than compensates for Hubble's modest 94-inch mirror. Hubble continues to churn out discoveries, ranging from dust storms on Mars to black holes in neighboring galaxies to objects at the very edges of space and time.

Today Hubble and all its Earth-based companions have been bested—reduced to the role of eyepieces for a "telescope" vastly greater in scale. This "telescope" consists of a cluster of some 10,000 galaxies, located 2 billion light-years from Earth, that acts as a vast cosmic lens. The gravity of this huge assemblage of matter bends light rays and thus concentrates light from more distant objects, making them visible to the paltry telescopes of our own making. In 2001, this cosmic telescope led to the discovery of a "baby galaxy" that formed some 13.4 billion years ago—only 600 million years after the Big Bang and in a realm of time astronomers call the "Dark Ages" because most stars had not yet formed. The cosmic telescope provides a 30-fold enhancement of the light from the baby galaxy; without this effect, the galaxy would be utterly invisible even to the most advanced human-made telescopes. By the way, the cosmic lensmaker didn't make this lens quite perfect; like other *gravitational lenses*, the huge galactic cluster that focuses light from the distant baby galaxy forms distorted and often even multiple images of objects that lie beyond it. In fact, it was through such multiple imaging that gravitational lenses were first identified.

A Common Absurdity

Time loops? A time before time? A many-branched Multiverse? Wormholes? Leapfrogging into the future? Galactic telescopes peering to the edge of space and time? What do all these seeming absurdities have in common? All require conceptions of space and time that boggle our common sense. All require that space and time bend, warp, and distort in ways that have no counterparts in our everyday experience. All require that we give up our tenth-grade notions of geometry, with straight lines and perfect triangles. Straight lines through space, from here to there? Not in a Universe where cosmic lenses focus multiple images of the same object! Straight lines through time, from now to then? Not in a Universe with wormholes, time loops, and high-speed star trips!

The geometry of space and time is not, in fact, the geometry you learned in tenth grade. It's a much richer geometry that allows for the curving, bending, or warping of space and time. That richness, in turn, enables a host of new phenomena: time loops that bend time back on itself; black holes, where spacetime curvature becomes infinite; gravitational lenses that create multiple paths for light from distant sources to reach Earth; high-speed travel that's a shortcut to the future. The strange geometry of spacetime is not some unfathomable mystery, though. It's described precisely by Einstein's theory of relativity—a theory that is as much about the geometry of space and time as it is about phenomena of physics. In fact, relativity suggests that we think of geometry as a branch of physics rather than mathematics, setting the spacetime "stage" on which we and the rest of physical reality strut and fret our roles in the larger Universe.

So what is this relativity theory that mixes geometry into physical reality and leads to mind-boggling possibilities like time loops, wormholes, and cosmic telescopes? Ultimately, the idea behind relativity is a very simple one—so simple that I can state it in a single sentence. But it's a very big idea, too—an idea whose consequences and philosophical implications go far beyond the confines of physics. In the next chapter I will introduce you to the simple Principle of Relativity, and I'll convince you that you already grasp and even embrace this idea at the heart of relativity. From there we'll explore how relativity came to be, what it really means, and why its consequences fly in the face of common sense. At the end, I'll return to the cosmic implications of relativity to show how exotic phenomena like wormholes, time loops, and gravitational lenses follow from your newfound understanding of Einstein's remarkable vision.

TENNIS, TEA, AND TIME TRAVEL

• • •

Picture this: You're on a cruise ship, sailing straight and steady through calm waters. You're playing tennis on the ship's indoor court. How does the moving ship affect your game? As the ball approaches your racket, do you sense the ship's motion and adjust your swing accordingly? And what about your opponent, who has her back to the ship's motion? Does she have to adjust her play in a way different from you?

The answer to all these questions is an obvious no. Your game on the cruise ship proceeds exactly as it would on land. Neither you nor your opponent needs to consider the ship's motion whatsoever. Neither of you needs to adjust your playing style to compensate for that motion.

Here's something simpler than the complex volleying of a tennis match. Suppose you're standing on land, with no wind blowing, and you throw a tennis ball straight up. Up the ball goes; eventually it slows to a momentary stop, then drops straight back down into your waiting hand. Now try the same thing in that indoor court on the cruise ship. (I've put the court indoors to eliminate the wind you would experience on an outdoor court because of the ship's motion through the air.) Again the ball goes straight up and drops straight back down. That the ship is moving doesn't matter in the least.

Tired of tennis, you stroll over to a microwave oven to heat a cup of tea. Now how do you compensate for the ship's motion? Do you put the cup at the sternward end of the oven because the microwaves get "left behind" by the moving ship? Do you adjust the power level because the microwaves behave differently from how they would back on shore? Of course not! The oven, like the tennis ball, behaves in just the same way on the moving ship as it would on solid ground.

Of course, a cruise ship doesn't go very fast—maybe 30 miles per hour or so. So imagine now that you're in a space colony on Venus, which at that point in its orbit happens to be moving about 30 miles per *second* with respect to Earth. The atmosphere inside the colony is the same as on Earth, and Venus's gravity has essentially the same strength as Earth's. You step into the colony's recreation center for a game of tennis. As you start your serve, do you need to account for the fact that the whole court is moving at 30 miles per second? How could you possibly compensate for that speed? The answer, of course, is that you don't have to. The tennis game on Venus proceeds just as it would on Earth. Venus's motion is irrelevant.

Now try the simpler act of throwing the ball straight up. One second later, is it 30 miles away from you? Of course not! It goes straight up and comes straight down, just as it would on Earth. And why not? Why should the existence of Earth, millions of miles away and moving at 30 miles per second with respect to Venus, have any significant effect on what happens on Venus? Why should the rules governing the tennis ball be any different on Venus than they are on Earth?

Tired of Venusian tennis, you stroll over to a microwave oven to heat a cup of tea. This oven, of course, is moving at 30 miles per second. Do you worry about that? Of course not! Again, Venus's motion is irrelevant.

Maybe even 30 miles per second isn't much. So now imagine you're a humanoid being on an Earthlike planet in a distant galaxy moving away from Earth at 80 percent of the speed of light—about 150,000 miles per second. (The Hubble Space Telescope routinely observes very distant galaxies with speeds this high; their motion is part of the overall expansion of the Universe.) Again, it's tennis time. As you start your game, do you think about the fact that you're moving at 150,000 miles per second away from Earth, and do you make a hopeless attempt to compensate for that colossal motion? Of course not! You don't even know about Earth; the Milky Way galaxy, in which Sun and Earth are minor players, is but a faint smudge in your civilization's most advanced telescopes. Why should Earth have anything to do with you and how you play tennis?

Again, the game's over and it's tea time. The microwaves that heat your tea move through the oven at the speed of light—and this time the oven itself is moving at 80 percent of that speed. Do the microwaves get left behind? Does the microwave oven need a different set of instructions—or a whole different design—to work on this distant planet because it's moving at such high speed? The answer, again, is a firm no. The microwave oven works just fine, thank you, on this distant planet as it does on Earth.

These three examples of the cruise ship, Venus, and the distant galaxy all illustrate a point that seems to make perfect sense: The physical "stuff" of the Universe (tennis balls, atoms, microwaves—indeed, all matter and energy) works the same way everywhere. Put another way, the laws of physics—the rules that are somehow written into the structure of the Universe and that tell things how to behave—are the same everywhere. There's no special, preferred place where the laws of physics are correct while they need modifying everywhere else.

This idea that no place is special is hardly new. In 1543, Copernicus's *On the Revolutions of the Heavens* first challenged the ancient view of Earth as the center of the Universe. Copernicus's Sun-centered theory was a major shift not only for science but also for philosophy and religion (which is why the established church fought Copernicus's ideas). Since Copernicus, discoveries in science have only reinforced the view that there's nothing special about Earth in the cosmic scheme of things. We, its inhabitants, may love our planet and consider it a special place—but that doesn't mean Earth gets special privileges where the laws of physics are concerned.

Actually, "place" isn't quite the right word here. While it's true that the laws of physics apply everywhere, more important in my tennis examples is that they apply *in different states of motion*. A tennis match and a microwave oven—both governed by the laws of physics—work the same way on Earth and on a hypothetical planet moving away from Earth at 80 percent of the speed of light. The fact of the planet's motion is simply irrelevant. If it weren't irrelevant, there would be something mighty special about Earth, in that everyone else in the Universe would have to reference Earth in describing their own physical situations.

This idea that motion doesn't matter in some absolute sense goes back to Galileo and Isaac Newton, who were first to formulate quantitative laws that describe the way objects move. But the idea found its complete fruition with Albert Einstein, whose theory of relativity is based on the simple statement that motion doesn't matter, that is, that the laws governing physical reality are just the same on that distant planet hurtling away from Earth as they are right here on Earth itself. I'll rephrase that absurdly simple fact: The theory of relativity is, in its barest essence, just the simple statement that regardless of one's state of motion the laws of physics are the same.

If you followed the simple example of the tennis match and the microwave oven, then you already know and believe the basis of Einstein's theory of relativity. At its heart, this really is the essence of relativity: that you can play tennis and use a microwave oven— or undertake any other activity involving physical reality—in any state of motion, and you will always get exactly the same results. Put in more scientific language, the results of any scientific experiment will be the same for anyone who cares to undertake that experiment—again, regardless of state of motion. By "experiment," I simply mean some activity that probes the behavior of physical reality. In that sense, playing tennis is an experiment. So is heating water in a microwave oven. Watching the tennis ball bounce or the water boil tells you about how the physical world works. A simpler and more controlled experiment is the third one I introduced: throwing a ball straight up and observing its subsequent behavior. What relativity says about these and any other scientific experiments is that otherwise identical experiments performed by experimenters in different states of motion will give the same results. By "identical" I mean all relevant physical circumstances are the same. That's why I put the tennis court on Venus and on an Earthlike planet in a distant galaxy; that way, gravity was the same and so, therefore, was the force with which you had to swing the racket to put the ball in your opponent's court. On Jupiter or the Moon your tennis game would go a bit differently because of the stronger or weaker gravity, but ultimately you could infer the same underlying physical laws. Stated more directly, anyone who cares to experiment with physical reality will discover the same fundamental physical laws, regardless of the experimenter's motion.

So you already know and believe the essence of relativity. In that sense, you can close this book and be done with it. Stripped to its essentials, the theory of relativity is summarized in a single, simple English sentence: The laws of physics are the same for all. This statement—called the *Principle of Relativity*—is so obvious that you need read no further, so obvious that I've probably belabored it too much already.

But wait! Although the essential idea of relativity is simple, straightforward, and so eminently believable as to be obvious, the consequences of that idea are anything but transparent. Nor, at first glance, are those consequences particularly believable. I gave you some examples in the preceding chapter; they included such seeming absurdities as time loops, wormholes, and shortcuts to the future.

Consider the last of these, in which you and I, originally the same age, find ourselves 20 years apart when you return from a trip to a nearby star. Is this really possible? On what absurd new principles of physics does it rest? It is possible and was verified, albeit modestly, by the round-the-world atomic-clock experiment I described. For high-speed subatomic particles, the observed effect is even more dramatic than for your hypothetical star trip. But there's no bizarre new principle involved—only the principle that the laws of physics should be the same regardless of one's state of motion.

Common Sense or Uncommon Sense?

The story of your jump into the future is hard to swallow because it seems to violate everything your common sense tells you about the nature of time and space. Well, your common sense is wrong! Without going into the details of just how it's wrong (they'll come out in subsequent chapters) I'll simply say that your common sense is wrong because it's based on a very limited experience that masks the full richness that the physical Universe has to offer. No matter how cosmopolitan, no matter how cultured and well read, and no matter how much of a world traveler you are, it's a simple fact—no offense intended—that you're very limited and provincial in your outlook. Why? Because you've spent all your life on a small, cool chunk of rock called Earth. Unless you're an astronaut, a military pilot, or have flown on the Concorde, the fastest you've ever moved *relative to things that are significant to you* is about 600 miles per hour in a jet airplane. That italicized phrase is important, because you do move at much greater speeds relative to some other things. For example, you move at some 30 percent of the speed of light relative to the electrons that beam through your TV to create the picture you see on the screen. But you have no direct experience of those electrons, so they play no role in establishing your commonsense notions. And you're moving at more than 90 percent of the speed of light relative to the most distant galaxies observed to date—but again, you don't experience those galaxies in any significant way, so they play no role in developing your commonsense ideas. Were you an engineer designing TV picture tubes or an astrophysicist studying distant galaxies, it would be obvious to you that something isn't quite right with your commonsense notions of time and space.

Think a minute about how those notions developed. As a baby, you crawled about the house, maybe tumbled down the stairs, and gradually formed your concept of three-dimensional space. Later you walked, ran, bicycled, drove, or flew—and all those activities only reinforced your childhood understanding of space as a kind of absolute stage for the activities of your life. As a young child on a long car trip, your plaintive "When will we be there?" showed an embryonic conception of time. That, too, grew into your adult notion of time as an absolute, universal measure of progress from past toward future. But again, that sense of an absolute space and time—with you and your friends occupying the same spatial stage and aging at the same rate—grew out of your very limited experience on this one planet. As far as concepts of space and time are concerned, the particularly limiting aspect of your experience is your rather slow movement in relation to anything significant to you.

If you, as a baby, had crawled about at nearly the speed of light—again, relative to important things like your parents or the household furniture—then you would have no need of this book, and the world would not have needed Einstein. Relativity and its consequences would be obvious to you and having a friend jump into the future would be entirely consistent with your commonsense notions of time and space.

If you accept the point of this chapter—that the laws describing physical reality should be the same for everyone—then the basic principle behind relativity already makes sense to you. But consequences that follow logically from that principle seem to make no sense—like that 20-year gap between our ages following your star trip. In the remainder of this book I want to show you just why we're so sure of the Principle of Relativity and how that principle leads inexorably to conclusions that seem to fly in the face of common sense. At the other end, I hope you'll come to accept those conclusions and the phenomena they entail as manifestations of a Universe whose richness is far greater than your everyday experience would suggest.

MOVING HEAVEN AND EARTH

• • •

I've just convinced you that the laws governing physical reality should be the same on Earth, on Venus, and on a planet in a distant galaxy moving away from Earth at nearly the speed of light. My task was easy because you already knew that Earth is not the center of the Universe and you accept that astrophysicists routinely observe distant galaxies moving rapidly away from us. If the laws of physics apply throughout the Universe, surely they should work as well in those distant galaxies as they do here—especially given that neither we nor denizens of distant galaxies have any claim to centrality in the grand scheme of things. So you came away from the preceding chapter realizing that you already understand Einstein's Principle of Relativity—the simple statement that the laws of physics don't depend on your state of motion.

After showing that you already believe the basic premise of relativity, I reviewed the strange tale of the star trip that could leapfrog you into the future. That was disturbing because it violated what your common sense tells you about the nature of time, and it was but a glimpse of other disturbing things to come. So, although the idea behind relativity is simple, its consequences are not. I could plunge further into those consequences right now, but you'll find it more convincing if we first explore how the concept of relativity arose and why it required a genius of Einstein's stature to assert what is, in its essence, such a simple idea.

In the next few chapters I'll back up and give a brief historical overview of the evolution of our understanding of physical reality. I'll show how the idea of relativity became inevitable, but not until the early years of the twentieth century. As you contemplate this history, remember that you have the advantage of knowing a lot more than even the most advanced scientists of earlier times. Pre-Copernican scientists lacked your broader perspective on a Universe in which Earth's place is not special. It was only around 1920, when Einstein's relativity was already becoming widely accepted, that astronomers recognized the existence of other galaxies. And not until the last quarter of the twentieth century did instruments like the Hubble Space Telescope make possible the discovery of distant galaxies moving relative to us at speeds approaching that of light. The notion of a tennis match in such a galaxy—a notion that helped convince you of the obviousness of the relativity principle—would have had little credibility much before your time.

A Matter of Motion

The history of physical science is intimately connected with our understanding of the nature of motion. If you ever took a high-school physics course, you studied motion—a subject you may not have found particularly exciting. But the study of motion is profound, for several reasons. First, motion is the source of all change. Imagine a world without motion: Earth stops rotating, so it's perpetual daytime. Earth stops revolving around the Sun, so it's always the same season. Your body can't move, so you're stuck forever in one spot. Atoms cease moving, so there's no chemistry—no release of energy, no change in the substances of the world. Nothing evolves, transforms, mutates, develops, or otherwise changes. And there are no molecules jumping across the synapses of your brain, so there's no thought. Absent motion, *everything* stops. Period.

Conversely, if motion does exist (as it obviously does), then understanding motion will help you understand night and day, the seasons, the chemical reactions that result from the motion of atoms and the electrons within them, and even the functioning of your brain as it's based in the motion of molecules. Understanding motion at the most fundamental level goes a long way toward explaining the behavior of matter.

Furthermore, you've already seen hints that relativity is going to do strange things to your commonsense concepts of time and space. Those concepts are themselves closely associated with the idea of motion. What does it mean to move? It means getting from one place to another, and doing so in some time. Whatever else motion means, it involves passing *through time* and *through space*. So motion holds the key to understanding time and space.

Two Questions about Motion

What causes motion? That seems an obvious question and one that probably suggests some obvious answers. Push a chair, and it moves across the floor. Stop pushing, and it stops. Push your car, by stepping on the gas and letting the engine and tires do the actual pushing, and your car goes. Take your foot off the gas, and the car soon stops. Drop a ball, and Earth's gravity pulls on it, causing it to move. Pushes and pulls seem to be the causes of motion. The physics word for pushes and pulls is *force*, a word that I'll use interchangeably with push or pull. Some forces are obvious, like the direct push of my hands on that chair; others, like the pull of Earth's gravity or the repulsion of two magnets, are mysteriously invisible.

To the ancient Greeks, especially Aristotle (384–322 BCE) and his followers, the idea that force causes motion was obvious. An ox pulled an oxcart, making it move. Stop the ox and the cart stops. Other examples took a little more explaining. An arrow, for example, whizzes through the air without an obvious push. Aristotelians explained this motion by saying that air rushes from ahead of the arrow to behind it, exerting the force that keeps it moving. So for Aristotelians the answer to our question was obvious and similar to the answer you would probably give: force causes motion.

There's a more subtle question we can ask about motion: Is there a natural state of motion? By a natural state, I mean a state of motion that requires no explanation—a state that an object naturally assumes unless something is explicitly done to it, like pushing or pulling it. For the Aristotelians, the answer to this second question is implicit in the answer to the first. Since it takes a force—a push or pull—to sustain motion, the natural state must be the state of rest. If you see something at rest, that's expected; it requires no explanation. But if you see something moving, you want an explanation, an answer to the question, What's making it move?

For the ancient Greeks our questions about motion admit not one but two answers. That's because the ancients distinguished two distinct realms of physical reality, the terrestrial and the celestial. It's in the terrestrial realm that the natural state is to be at rest—meaning, given the terrestrial context, being at rest with respect to Earth. In fact, the Greeks' idea of motion's natural state incorporates also a simple explanation of gravity: the natural state for objects in the terrestrial realm is to be at rest as close as possible to the center of Earth, which itself was the center of the Universe.

The heavens, in contrast to the base, ordinary realm of Earth, are the abode of the gods and the realm of perfection. Here the stars and planets move naturally in the most perfect of paths, namely, circles centered on Earth. So the natural state of motion in the celestial realm is circular motion—a state that is so obvious as to require no further commentary.

There's a slight problem with this picture of simple circular motion throughout the heavenly realm, in that it can't quite explain the subtle motions of the planets. Observed night after night, the planets seem not to go uniformly in the same direction as would be the case if they simply circled Earth, but on occasion they backtrack in what astronomers call *retrograde motion*. In the second century CE, the astronomer Ptolemy solved this problem by positing that the planets actually move in smaller circles (called *epicycles*) carried around on larger circles that are themselves centered on Earth. By adjusting the size and speed of the motion on these smaller circles, Ptolemy was able to match closely the observed planetary motions. Figure 3.1a shows a simplified view of the Ptolemaic scheme of the Universe. Ptolemy's epicycles helped preserve two essential philosophical bases of ancient science: (1) Earth is at the center of the Universe, and (2) perfection reigns in the celestial realm, here in the form of motion in perfect circles.

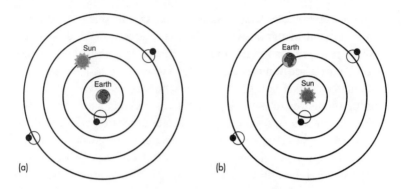

Fig. 3.1 Simplified views of (a) Ptolemy's Earth-centered universe and (b) Copernicus's heliocentric system. In both systems all motions in the heavens involve perfect circles, which for the planets (black dots) become small circles about larger circles. Neither diagram is to scale and both omit some details.

Heresy!

Among the most profound advances in all of science is Polish astronomer Nicolaus Copernicus's suggestion that the Sun, rather than Earth, is the center of the Universe, the point about which all other celestial bodies revolve (Figure 3.1b). Copernicus (1473–1543) first hit on this idea early in the sixteenth century and expounded it fully with the 1543 publication of his *De revolutionibus orbium coelestium libri vi* ("Six Books Concerning the Revolutions of the Heavenly Orbs"). This new idea—heretical because established church dogma was rigorously Aristotelian—was the first in a series of intellectual steps that stripped Earth of its special place in the grand scheme of things. Einstein's relativity, as I outlined it in the preceding chapter, is the ultimate step in that series.

Copernicus's heliocentric proposal was truly revolutionary, but in other ways his new model for the Universe clung conservatively to older ideas. Although he removed Earth from its central place, Copernicus nevertheless maintained the distinction between terrestrial and celestial realms. For him, the heavens remained the realm of perfection. Sun, planets, and stars were perfect bodies, and they moved, appropriately, in perfect circles. Having Earth and the other planets in circular paths around the Sun helped explain many astronomical observations that had puzzled the ancients. In particular, the retrograde motion of the planets and their apparent brightening and dimming over time both followed naturally from the rather complicated path a Sun-orbiting planet describes when viewed from a Sun-orbiting Earth. Copernicus's placement of Earth among the other planets raised the question of how the terrestrial realm, with its imperfections and pestilences, could be part of the heavens. Nevertheless, Copernicus maintained the terrestrial/celestial distinction and did not change the Aristotelian answer to the question, Is there a natural state of motion? On Earth, the Copernican answer remained "at rest" and in the heavens it remained "motion in perfect circles."

Copernicus's ideas inspired the Danish astronomer Tycho Brahe (1546–1601) to undertake extensive, regular, and highly accurate observations of stars and planets. Tycho ultimately rejected both the Ptolemaic and Copernican views in favor of his own more complicated Earth-centered model. When Tycho died, he left his prolific set of astronomical data to his assistant, the German astronomer Johannes Kepler (1571–1630). Working especially with Tycho's data for Mars, Kepler conceived a radically simple way to avoid the complex combinations of circles needed in the models of Ptolemy, Copernicus, and Brahe. Compounding the Copernican heresy, Kepler dispensed with the notion that celestial bodies must move in perfect circles. Instead, he showed, the simplest explanation for planetary observations was that the planets move in elliptical orbits, with the Sun not at the center but at a special point called the focus of the ellipse (see Figure 3.2). And instead of incorporeal spirits pushing planets in their circular paths about a passive Sun, Kepler proposed that the Sun itself exerts the push that gives the planets their elliptical motion. Furthermore, Kepler succeeded in formulating quantitative laws that describe how the speed of the planets varies as they move about the Sun. Although Kepler *described* planetary motions accurately, his theory couldn't *explain* them. Philosophically, though, he had taken a big step—freeing the celestial realm from the constraint of perfectly circular motion.

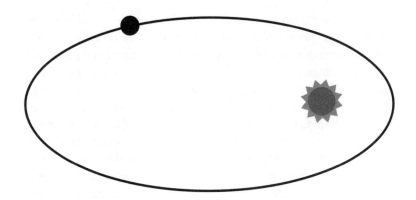

Fig. 3.2 Kepler showed that each planet moves in an elliptical orbit, with the Sun at a special point called the focus. The ellipse shown here is highly exaggerated; most planets' elliptical orbits are nearly but not quite circular.

Celestial Acne and Other Galilean Heresies

While Kepler struggled to interpret Tycho's voluminous astronomical data, his contemporary, the Italian Galileo Galilei (1564–1642), was also reading "the book of nature." Galileo insisted on the primacy of experiment and observation in establishing scientific truth, moving science from what had been essentially a branch of philosophy to an empirical and precisely mathematical description of physical reality.

In 1609, Galileo constructed the first astronomical telescope. With this new tool, he soon found convincing evidence supporting both Copernicus's heliocentric Universe and Kepler's abandonment of celestial perfection. Turning his telescope on Jupiter, Galileo discovered the four largest of the many moons that orbit the giant planet. Now known as the Galilean satellites, these four miniature worlds circle Jupiter with periods of a few days. As you can verify with binoculars, their position thus changes nightly. Often fewer than four are visible, as they pass behind or in front of Jupiter itself. To Galileo, the Jovian system was obviously a miniature version of Copernicus's heliocentric solar system, with Jupiter playing the role of the Sun and its moons, the planets. If Jupiter can host such a system of satellites, reasoned Galileo, why not the Sun?

Galileo's support of Copernicus's heliocentrism strengthened further with his telescopic observations of Venus. Like Earth's moon, Venus exhibits phases that result from our seeing different portions of the planet illuminated with sunlight. Furthermore, when Venus is full it appears smallest, and when it's a narrow crescent it appears largest. That's not easy to explain with an Earth-centered model in which Venus remains the same distance from Earth. But it follows naturally in the heliocentric model, since Venus needs to be on the other side of the Sun—as far as possible from Earth and thus appearing smallest—for us to see it fully illuminated.

When Galileo turned his telescope on the Moon, he discovered its rugged, mountainous terrain—making the Moon a body much more Earthlike than the purportedly perfect spheres of the celestial realm. Galileo also became one of the first Europeans to study sunspots—those dark blotches that we now know as cooler regions of intense magnetism on the solar surface. (The Chinese had discovered sunspots much earlier and called them "crows," but it wasn't clear that they were actually features on the Sun itself.) Like the cratered, mountainous Moon, the blemished Sun gave evidence for imperfection in the celestial realm.

Alas for Galileo, his scientific efforts on behalf of Copernicus's ideas won him no favor with those in power. For Galileo's staunch advocacy of the heliocentric heresy, the Inquisition in 1633 sentenced him to life imprisonment, and he spent the rest of his days under house arrest. Repeated attempts failed to clear his name and Galileo's *Dialogue Concerning the Two Chief World Systems, Ptolemaic & Copernican* was on the Vatican's banned book list until 1835. Not until 1992 did the Pope issue a statement essentially exonerating Galileo.

Galileo explored not only the heavens but also the terrestrial realm. Legend has it that he dropped stones of different masses off the Leaning Tower of Pisa and observed that they hit the ground at the same time. Whether or not the legend is true, Galileo did argue on logical grounds that all objects must fall at the same rate. To study gravity further, he rolled balls down inclines so as to dilute the

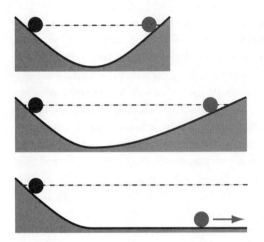

Fig. 3.3 This "thought experiment" led Galileo to the conclusion that a moving object should continue forever in uniform motion unless a force acts to change its motion.

effect of gravity and thus permit more detailed observation. Galileo noticed that a ball rolling in a trough-shaped incline rose to the same height it had started from. On a more gradual incline, the ball would roll farther horizontally before reaching its starting height. From this Galileo concluded that the ball would roll forever on a purely horizontal plane (see Figure 3.3).

Galileo's conclusion represents a dramatic shift in our understanding of motion, for it suggests that motion itself needs no cause or explanation. Specifically, an object moving horizontally in a straight line with constant speed will continue to do so unless something acts explicitly to slow it down, speed it up, or change its direction. For Galileo, our two questions about motion have new answers. What is the natural state of motion? For Galileo, it's no longer rest, but motion at constant speed in a straight line. That state—called *uniform motion*—requires no explanation. And what causes motion? For Galileo, that's no longer the right question. Motion itself is natural, requiring no cause or explanation—at least if it's uniform motion, unchanging in speed or direction. What does require explanation is any *change* in motion. And what causes such change? A force, that is, a push or a pull. So for Galileo, forces don't cause motion; rather, they cause *changes* in motion.

Motivation for Galileo's work on terrestrial motion came from a problem inherent in the Copernican scheme. If Earth moves around the Sun, why don't we feel Earth's motion? Why aren't we left behind or, since we're not, why don't we feel a push or pull to keep us moving along with Earth? Galileo's new understanding of motion provides the answer. Since we and everything else on Earth partake of our planet's motion, we continue naturally to move with the Earth—a state of motion that requires no further explanation. (Earth's motion isn't exactly uniform, since the planet is both rotating on its axis and revolving about the Sun, but this is a minor complication to an essentially solid argument.) So we don't feel Earth's motion because we share it and are therefore at rest relative to Earth, even though in the Copernican scheme we and Earth are "really" moving.

The Genius of Cambridge

The year 1642 saw the death of Galileo and the birth of Isaac Newton, whose genius was to dominate physics until Einstein's time. After a decidedly rough childhood, Newton enrolled at Cambridge University in 1661. But in 1665, just after Newton received his bachelor's degree, the plague swept across Europe. The universities closed and students scattered to the countryside. Newton returned to his native Woolsthorpe for a period of 2 years. But he was far from idle: during that time Newton developed much of his theory of light, explored the orbital motions of the Moon and planets, and laid the groundwork for his theory of gravity.

A popular tale has Newton sitting under an apple tree, being struck on the head by a falling apple, and thus discovering gravity. But it didn't take Newton's genius to recognize gravity in a falling apple. Our ancestors, swinging through the trees, already knew instinctively about gravity here on Earth. If there's a deeper truth to the apple tale, it's that Newton, while sitting under the tree, was contemplating the Moon. And his remarkable stroke of genius was this: Newton realized that the motion of the apple and the motion of the Moon are *the same motion*. Both are "falling" toward Earth, both under the influence of the same force, namely, Earth's gravity.

Newton coined the term "gravity" (from the Latin *gravitas*, or heavy) to describe this force, and he proposed that gravitation is a truly universal phenomenon. Every object in the Universe, Newton claimed, attracts every other. The force of attraction depends on the masses of the objects and the distance between them. Move two objects farther apart and the gravitational force between them weakens. Specifically, if you double the separation, the force drops to one-fourth of its original value; triple the separation, and it drops to one-ninth. That is, the force drops as the inverse square of the distance between the two objects. Newton verified this mathematical property by comparing the acceleration of the falling apple with that of the Moon, some 250,000 miles from Earth and thus subject to much weaker gravity.

If gravity acts between all objects, why don't everyday objects fall together in a great clump? The answer is that gravity is an extraordinarily weak force. Only large accumulations of matter—like planets and stars—are sufficiently massive to have obvious gravitational effects. For smaller objects, other forces usually overwhelm gravity. Nevertheless, it's possible with sensitive equipment to measure the gravitational force between objects as small as apple-sized lead spheres. And at the opposite scale, Newtonian gravitation accounts even for the slow, stately motion of the galaxies themselves. Such observations of gravity, from laboratory to cosmos, confirm the remarkable scope of Newton's ingenious idea.

Before exploring Newton's work further, let's pause to note the radical philosophical shift inherent in Newton's recognition of universal gravitation. Now, for the first time, the same laws describe motion on Earth and in the heavens. No longer are there two realms with two different sets of physical laws. Instead, a single physics governs the entire Universe, from everyday events on Earth to the most distant reaches of the cosmos. With universal gravitation, Newton has taken another giant step away from the notion that Earth is somehow special—a step that helps pave the way for relativity.

How is it that both apple and Moon are falling, when the apple is on a collision course with the ground while the Moon remains forever in its orbit? Both are falling in the sense that both are moving under the influence of gravity alone, and both are deviating from the straight-line, constant-speed motion that is the natural state in the absence of any force. Furthermore, for both apple and Moon the direction of the deviation is the same—toward Earth. Thus Earth's gravity exerts a force on both apple and Moon, and that force does what Galileo recognized that forces do: it causes *change* in motion.

The Moon's motion differs from the apple's in its specifics but not in the fundamental fact that both motions constitute "falling" under the influence of Earth's gravity. The difference between the two motions is that the apple drops, starting from rest, and thus falls straight down. The change in its motion consists entirely of an increase in speed. The Moon, in contrast, has a substantial speed in a direction at right angles to "down." Absent any force, the Moon would move in a straight line at constant speed. Earth's gravity changes that motion, pulling the Moon into the circular orbit that it follows repeatedly. The Moon is falling toward Earth in the sense that that's the direction of its deviation from what would be a straight-line path, but because of its sideways motion it never gets any closer to Earth.

Using the reasoning we've just applied to the Moon and the apple, Newton became the first to envision the possibility of artificial satellites. He imagined Earth with a high mountain and considered what would happen to objects launched horizontally from the mountaintop (Figure 3.4). Give the object just a little speed, and it falls to Earth, landing near the foot of the mountain. Give it more speed and it strikes Earth farther from the mountain. In each case, gravity pulls the object out of the straight-line path it would follow in the absence of any force. Give the object just the right speed and the deviation from the straight-line "natural" motion is just enough to follow Earth's curvature. At that point the object is in a circular orbit, always falling *toward* Earth but never getting any closer. That's essentially the situation of the Moon and of many artificial satellites. Throw the object still faster and its orbit becomes an ellipse—just as Kepler had described for planetary orbits.

Universal gravitation is only one of Newton's many contributions to physics. Building on Galileo's work, he made precise the idea that the natural state of motion is uniform motion—straight-line motion

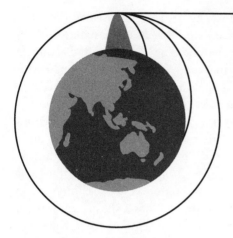

Fig. 3.4 Newton envisioned launching projectiles from a high mountain. The faster they're launched, the farther they go before hitting Earth. Fast enough, and a projectile "falls" around Earth in a closed orbit. Even in this orbit, the projectile is falling in the sense that it's deviating from the straight-line path (shown) that it would follow in the absence of Earth's gravity.

at constant speed—and that forces cause *changes* in motion. Newton's *first law* restates that point, declaring that an object at rest remains at rest and an object in uniform motion remains in uniform motion, unless it's subject to some force. His *second law* tells just how much change a given force causes. That change is characterized by the object's *acceleration*, or rate of change of motion. The acceleration, says Newton's second law, is larger for a larger force and smaller for an object of larger mass. If you've studied physics before, you've seen the second law expressed in the well-known equation *force = mass × acceleration*. For Newton, acceleration doesn't just mean speeding up. Rather, any *change* in motion—whether in speed, direction, or both—entails acceleration. The falling apple gains speed as it approaches Earth, and thus it's accelerating. Moon's speed doesn't change, but the direction of its motion does because it's describing a circular orbit. So its motion is changing, and thus it too is accelerating.

Newton's *third law* is probably familiar to you in the statement "for every action there is an equal and opposite reaction." Actually, that language obscures the third law's simple physical meaning. What the law really says is that if I push on something, then that thing pushes back on me with an equal force—equal, that is, in strength, but opposite in direction. More generally, forces always come in pairs: if A pushes on B, then B pushes back on A. Applied to gravity, the third law means that when Earth pulls on the falling apple, the apple pulls with equal force back on Earth. Because the apple's mass is small, Newton's second law shows that the apple gets a hefty acceleration. In contrast, Earth is enormously massive, so its acceleration is negligible. But in principle Earth does accelerate slightly toward the apple. Both forces—Earth on apple and apple on Earth—are needed for a full and consistent description of motion in Newton's physics.

Newton's work provided the basis for a full description of phenomena involving motion. Collectively, Newton's laws and all that follows from them compose the branch of physics called *mechanics*.

A Theory of Everything?

Newton's three laws of motion, coupled with his law of universal gravitation, seemed capable of explaining essentially all physical phenomena. With Newtonian mechanics as a foundation, physicists could understand not only motion in the cosmos but also such seemingly unrelated phenomena as the behavior of gases under conditions of changing temperature and pressure. Although at first glance gas behavior seems to have little to do with Newton's laws, in fact it follows directly by applying those laws to the motions of the individual molecules constituting the gas. So great was the triumph of Newtonian physics that science held the hope of explaining all physical phenomena ultimately in terms of the motions of particles obeying Newton's laws.

Galilean Relativity

At the heart of Galileo's and Newton's new understanding of motion is the idea of uniform motion as a natural state, needing no further explanation. It was this idea that helped Galileo accept Copernicus's moving Earth. Because everything on Earth shares a common motion and continues naturally in that state, Earth's motion is not obvious to us. Indeed, both Galileo and Newton knew that the laws of motion didn't favor a particular place or state of motion. Galileo reasoned, for example, that a rock dropped from a ship's mast would behave just the same way if the ship were moving steadily through calm water as it would if the ship were at rest. In both cases the rock would fall straight down and land at the base of the mast. We express this idea more generally with the *principle of Galilean relativity*, which says that the laws of motion work equally well for anything in uniform motion.

Your immediate surroundings—specifically, those things that participate in the same motion as you—constitute what physicists call your *frame of reference*. Earth is the frame of reference for someone at rest on the planet, and Galileo's ship constitutes the reference frame for someone doing experiments on the ship. Put in terms of reference frames, Galilean relativity becomes the statement that the laws of motion are valid in any uniformly moving frame of reference.

Galilean relativity should sound a lot like the idea I introduced in the preceding chapter: the laws of physics are the same, regardless of one's state of motion. That's essentially what Galilean relativity says, especially because, in Galileo's and Newton's time, many scientists believed that the laws of motion could explain all of physical reality. So a statement about the laws of motion becomes, to a Newtonian believer, a statement about all of physics. Here's the important point: Galileo and Newton, like you yourself, already understood and accepted the principle of relativity. For them, as for you, there was no conceivable experiment that could answer the question Am I moving? Galileo's rock dropped from a ship's mast provides an example: the rock always lands right at the base of the mast, so the outcome of the experiment can't be used to determine whether the ship is moving. Other questions, such as Is the ship moving relative to the shore?, Am I moving relative to Earth?, or Is Earth moving relative to the Sun? are answerable—but these are questions about *relative* motion. That's why the statement that the laws of physics are the same, regardless of one's state of motion, is a *relativity* principle. It implies that motion itself is undetectable and therefore meaningless; all that matters is *relative* motion. Ultimately, the relativity principle traces back to Galileo's recognition that motion—at least uniform motion—is a natural state that requires no explanation.

An everyday example should remind you how obvious and simple an idea is Galilean relativity. Imagine you're on an airplane, cruising through calm air at a steady 600 miles per hour. You're served a snack of airline peanuts. Do you need to think about the airplane's 600-mph motion in order to get the peanuts successfully to your mouth? Do you need to modify your understanding of the laws of motion to describe the peanuts' behavior in the "moving" airplane? Of course not! The airplane's "motion" is irrelevant. Newton's laws work as well on the plane as they do on the ground. That's why I put the words "moving" and "motion" in quotes. In the context of Newtonian physics, it makes no sense to consider that the airplane is "moving" and Earth isn't. The laws of physics work the same in both places, so neither has a claim on some special state of being at rest. If you want to claim that Earth is really at rest and the plane really moving, then I'll challenge you to come up with some physical test that will prove you correct. Pull down the window shades so you can't see Earth slipping past, wait for the plane to be cruising through calm air at constant speed, and then think up something you can do—that is, a physical experiment—whose results will be different in the plane than they are if you performed an identical experiment on Earth. You won't find it. That's what it means to say there's no experiment that can answer the question, Am I moving? By looking out the window you can justifiably assert that the plane is moving *relative to Earth*, but that's as far as you'll get. You're just as correct in asserting that Earth is moving relative to the plane. It simply doesn't matter; with Galilean relativity, there's no such thing as absolute motion or absolute rest. This is all just a rehash of the tennis-match argument from Chapter 2, where the universal scope of the argument made it even more obvious how absurd it would be to assert that Earth alone among all the cosmos is truly at rest.

Throughout this book I'll be making the point again and again that there's no experiment you can do to answer the question, Am I moving? or, equivalently, that identical experiments performed in different reference frames give exactly the same results. Let me make very clear what this does and does not mean. What it does mean is that if you and I, each in uniform motion but each moving relative to the other, set up and perform identical experiments, then we'll get exactly the same results. But those experiments need to be truly identical in all respects except for our relative motion. That's why I set the tennis match in Chapter 2 on an *indoor* court, to eliminate any wind that would result in an outdoor court from the ship's motion relative to the air. That's why Galileo's rock-drop experiment on the ship should really be in an enclosed space, or maybe behind a sail, to block the apparent wind caused by the ship's motion. And to be really careful, I should do the airplane experiment when the plane is at low altitude, to minimize changes in gravity with distance from Earth. In practice, it may be hard to set up circumstances that are identical except for relative motion, but in principle it's possible. And if we do, then identical experiments will give identical outcomes.

What my statement does not mean is that observers in relative motion must see the results of the *very same* individual experiment in the same way. If I watch Galileo's rock-drop experiment from onboard the ship, I'll see the rock fall straight down. If you watch the very same experiment from the shore, you'll see the rock following a curved path because it shares the ship's motion relative to you. But if we both perform *different* yet identically staged rock-drop experiments, you on shore and I on the ship, then we will get identical results—namely the rock falling straight down, taking exactly the same time as measured by our different but identical clocks, and so forth down to the last detail. We simply won't be able to use any of our results to decide which of us is moving and which isn't.

We can express the principle of Galilean relativity in many ways. "The laws of mechanics are the same for all observers in uniform motion" or "The laws of mechanics are the same in all uniformly moving frames of reference" are formal ways to state the principle. To say that "I am moving" and "I am at rest" are meaningless statements is another way of putting Galilean relativity. "You can eat dinner on an airplane" or "You can play tennis on a cruise ship" are statements of Galilean relativity applied to specific circumstances. "Uniform motion doesn't matter," "Uniform motion is undetectable," and "Absolute motion is a meaningless concept" are still other ways to state the principle of Galilean relativity.

There are two subtle distinctions between the Galilean relativity principle of this chapter and my more general introduction of the relativity principle in Chapter 2. First, Galilean relativity applies only to the laws of *motion*, whereas my earlier relativity principle was a statement about all of physics. To one who believes that motion can explain all of physics, there's no difference. But if we should discover new areas of physics that aren't based in the laws of motion, then we'll need to ask anew whether the relativity principle holds there as well.

The second distinction is that I left out the qualifying phrase "in uniform motion" in Chapter 2. But we definitely need that qualification in expressing Galilean relativity. The laws of motion are decidedly *not* valid if you're not in uniform motion. Try eating those peanuts when the plane encounters unexpected turbulence or while it's accelerating down the runway. Then the peanuts act in a decidedly non-Newtonian fashion. They don't stay put on your tray table. Try tossing one into your mouth and it goes violently astray, landing instead on the floor. Or try playing tennis with the cruise ship sailing through a hurricane. Your intuitive feel for how the ball is supposed to behave will fail you; more formally, Newton's laws of motion just don't hold in the nonuniformly moving reference frame of the storm-lashed ship. So at least as far as Galileo and Newton are concerned, statements about nonuniform motion, that is, *changing* motion—are meaningful. It does make sense to say "this airplane's motion is changing" even though it makes no sense, in the context of Galilean relativity, to say "this airplane is moving." Absolute motion is meaningless in the Galilean/Newtonian view, but *change* in motion is supremely meaningful and, indeed, is what Newton's laws of motion are all about.

As I move toward Einstein's *special* relativity in the next few chapters, I'll continue to maintain the distinction between uniform and nonuniform motion. Later, in Chapter 14, you'll see how Einstein's *general* relativity blurs that distinction. But the distinction between the laws of *motion* (with their associated Galilean relativity principle) and *all* the laws of physics will continue to play a major part as we develop first the special and then the general theory of relativity.

LET THERE BE LIGHT

•   •   •

Newton's laws are remarkably sweeping in their scope, explaining phenomena from the behavior of gases to the game of tennis to the motions of planets, stars, and galaxies. But are there branches of physics that don't fall under the umbrella of Newton's laws? What about phenomena like sound and light, our primary ways of communicating with the world around us? Or electricity, at the heart of modern technology and, more fundamentally, responsible for the structure and behavior of matter from atoms and molecules to our own bodies and brains?

Waves

We'll begin with sound, a classic example of a *wave*. What's a wave? Think of an ocean wave or the wave of standing people that sweeps around a sports stadium. In each case there's a disturbance of some *medium*. For ocean waves, that medium is the water. For stadium waves, the medium comprises the people in the stadium. The disturbance moves through the medium, temporarily upsetting the status quo and then moving on. But the medium itself doesn't go anywhere, although it may move briefly back and forth or up and down as the wave passes by. That is, water from the distant ocean doesn't actually move toward shore with the waves. Watch a boat in rough water: the boat bobs up and down as a wave passes, but it isn't carried shoreward with the wave (things get a bit more complicated in the shallow water where the wave breaks or as a surfer intentionally rides a wave, moving with the wave by sliding down its sloping front). Similarly, those sports fans rise from their seats, then return; they don't go anywhere, even as the wave circles the stadium.

So what does move in a wave? Certainly the disturbance itself—the displacement of water from its normal flat surface or the displacement of the sports fans' bodies from their sitting positions. In the process the wave carries *energy*. In our examples that energy is associated with the temporary lifting and movement of the water as the wave passes or with the standing people in the stadium. So a good definition of a wave is that it's a traveling disturbance that carries energy but not matter.

Two simple ideas seem inherent in our definition of a wave as a traveling disturbance—ideas that will play a key role in the development of relativity. First, since a wave is a disturbance, it seems logical that it must be a disturbance of *something*—and that something is what we've called the medium for the wave. Second, since a wave is a traveling disturbance, it must travel with some speed. What determines that speed? Simple: the properties of the medium. For water waves, speed depends on things like the density of water and the force of gravity. In fact, we can deduce that speed by applying Newton's laws to the motions of the water, taking account of forces like gravity and water pressure. So we can add water waves to our growing list of phenomena explained by Newton's laws.

Now, on to sound. This, too, is a wave, and one that we can also understand by using Newton's laws. A sound wave is a disturbance of the air, consisting of regions of high and low air pressure (Figure 4.1). The disturbance moves through the air, but the air itself only moves back and forth, and very slightly at that. The medium for sound waves is clearly the air. We can directly measure the speed of sound or we can calculate it by applying Newton's laws and accounting for the force associated with the sound wave's air pressure variations. Either way, the answer for air under normal conditions is about 700 miles per hour. That, incidentally, is also about 1,000 feet per second, or one-fifth of a mile per second, which is

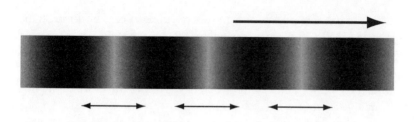

Fig. 4.1 A sound wave consists of alternating regions of compressed (dark) and rarefied (light) air. The entire structure moves with the speed of sound (large arrow). The air itself, however, only moves back and forth as the wave passes (small arrows).

why you count the time between lightning and thunder and divide by five to get the distance to the lightning strike. Each 5 seconds represents 5,000 feet or about a mile of travel for the sound waves. Implicit in this rule of thumb is an assumption that will soon become crucially important to us: the speed of light from the lightning is so fast that the travel time for the light is negligible.

So we understand sound; it's a wave ultimately governed by Newton's laws applied to the molecules that make up the air. The medium for sound is air, and the speed of sound is about 700 miles per hour. Seven hundred miles per hour relative to what? That's obvious—relative to the air, which is the medium in which a sound wave is a disturbance. Let me belabor this obvious point, since we'll soon be raising the analogous question for light. The speed of sound is 700 miles per hour relative to the air. If you're running through the air at, say, 10 miles per hour, then the speed of sound waves *relative to you* should be 710 miles per hour if you're running toward the source of the waves and 690 miles per hour if you're running the other way. No big mystery here; sound is a disturbance of the air, so to say its speed is 700 miles per hour is to say that sound moves at 700 miles per hour relative to the air. Sound must move at a different speed relative to someone who is moving through the air.

I'll finish with sound by returning to the question that began this chapter: Are there branches of physics that don't fall under the umbrella of Newton's laws? Sound most definitely is under the Newtonian umbrella, because it's entirely explainable by applying

Newton's laws of motion to air molecules. So sound is solidly a part of Newtonian physics, and we still don't have an affirmative answer to our question.

Light: Waves or Particles?

In *Opticks*, published in 1704, Newton himself gave one of the first thorough accounts of the nature of light. Inspired perhaps by the success of his laws of motion, Newton imagined light to consist of particles whose behavior—including the phenomena of reflection and refraction (the bending of light's path as it enters a transparent substance)—could be understood in terms of the laws of motion. Newton explained color, an obviously important property of light, by positing that different light particles excite vibrations that give us the sensation of different colors.

But Newton's *Opticks* had competition. In 1678 the Dutch scientist Christiaan Huygens proposed that light, like sound, consists of waves. Although Huygens did not address color, a wave theory of light provides a natural explanation of color in the different frequencies (rates of vibration) of the waves—just as different frequencies of sound waves correspond to different pitches. And although the wave and particle explanations were very different, no experiment at the time could distinguish them. However, Newton's stature was such that his particle theory dominated over Huygens's waves until the nineteenth century.

There's one thing waves do that particles cannot. Two waves can pass through the same place at the same time; when they do, the disturbances that each represents simply add to make a combined wave. This phenomenon is called *interference*. When crests of one wave meet crests of another, the two waves reinforce to make a larger wave. This is *constructive interference*. When crests meet troughs, the interference is *destructive* and the waves cancel. Figure 4.2 shows constructive and destructive interference. If the waves don't line up perfectly crest-to-crest or crest-to-trough, then the interference is somewhere between fully constructive and fully destructive.

In 1801 the English physician and physicist Thomas Young per-

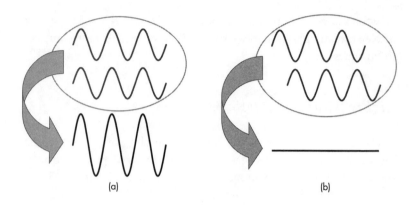

(a)   (b)

Fig. 4.2 Wave interference. (a) Constructive interference occurs when crests from two waves meet. (b) Destructive interference occurs when crests meet troughs.

formed an experiment that led ultimately to the downfall of Newton's particle theory of light. Young let sunlight pass into a darkened room through two pinholes, illuminating a screen opposite the holes. If light consisted of particles, one would expect to find two bright spots on the screen opposite the holes (Figure 4.3a). Instead, Young saw a series of alternating bright and dark spots on the screen. It's difficult to imagine how beams of particles could produce such a pattern, but the wave theory provides an obvious explanation (Figure 4.3b). Each hole acts like a source of waves, whose circular crests spread outward from the holes much like ripples from a rock dropped into a pond. The waves from the two holes meet and interfere. In some regions, crests meet crests and troughs meet troughs; here the interference is constructive and the wave strengthens. In optical terms, that means brighter light. In other regions, crests and troughs meet. Here the interference is destructive and the wave is diminished. Figure 4.3b shows that these regions of constructive and destructive interference lie along lines radiating from the vicinity of the two holes. Where the lines of constructive interference meet the screen, bright spots appear. At the lines of destructive interference, the screen is dark.

What determines where the bright and dark spots appear? That

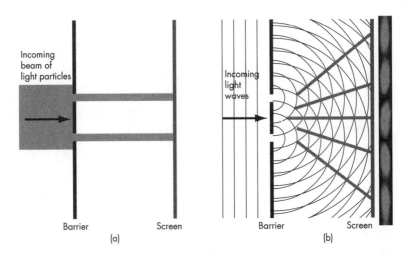

Fig. 4.3 Shining light through a pair of small holes in an opaque barrier would produce very different effects, depending on whether light consists of particles or waves. (a) Particles would produce two bright spots on the screen opposite the barrier. (b) Waves pass through the holes, producing circular waves, which subsequently interfere. Thick lines mark regions where crests meet crests, and troughs meet troughs. Here the interference is constructive, making the light brighter. Bright spots appear where these regions hit the screen. Between, destructive interference results in dark spots. To the right is an actual photo of the resulting pattern that appears on the screen.

depends on the spacing of the two holes and on the *wavelength* of the light—the distance between wave crests. The central bright spot, for example, forms from waves that have traveled equal distances from the two holes and thus meet "in step," with crests joining crests and troughs joining troughs. The dark zones on either side are from waves that have traveled further from one of the holes by just the right amount that crests meet troughs, making for destructive interference. Move the holes closer together and the bright and dark patches move farther apart, making the interference pattern more obvious. Move the holes farther apart, and the bright and dark patches get so close that they soon blur together. The reason the wave nature of light isn't obvious to us is that the wavelength of light is very small—around 20 millionths of an inch—so it takes very closely spaced holes to make Young's interference pattern noticeable.

So now we have an answer to our question about the nature of light: Waves or particles? That answer, clear from Young's interference experiment, is "waves." But waves of what? To answer that question takes us on a new and unexpected turn.

Electricity and Magnetism

The ancient Greeks knew of naturally occurring magnets and their attraction for iron, and they also studied the attraction of amber for small bits of cloth, straw, and similar materials—an attraction we now know to be electrical in nature. The Greeks even speculated on a relation between these two mysterious forces. But it was not until the late sixteenth century that fruitful study of electrical and magnetic phenomena began. That study culminated in the mid nineteenth century with a complete theory of electromagnetism that then led directly to Einstein's theory of relativity.

Experimentation with so-called static electricity revealed a new and fundamental property of matter, *electric charge*. It was Benjamin Franklin who clarified that there are two types of charge, which he called *positive* and *negative*. By 1788 the French scientist Charles Augustin Coulomb had expressed quantitatively the force between electric charges: like charges repel and opposites attract, with a force that depends on the charges and—as with Newton's gravitational force—on the inverse square of the distance between the charges.

Similar experiments with magnetism established that every magnet has a north and a south pole and that like poles repel while opposite poles attract. In contrast to electricity, no one has ever found an isolated north or south pole, which would be the magnetic analog of electric charge.

Meanwhile electrical science had advanced with the development of the battery and the subsequent study of electric current—the flow of electric charge in wires and electrically conducting solutions like salt water. Centuries of hunches that electricity and magnetism might be related culminated dramatically in 1820, with the Dane Hans Christian Oersted's discovery that an electric current deflected a nearby magnetic compass. The French physicists Jean-Baptiste Biot, Félix Savart, and André-Marie Ampère soon quantified the relation between electric current and magnetism.

Establishing the relation between electric current and magnetism marks the beginning of one of the great syntheses in human intellectual history—the joining of all electrical and magnetic phenomena, formerly believed distinct, under the single umbrella of electromagnetism. The hope for similar syntheses is a deeply philosophical driving force behind much of physics. Today, that hope manifests itself in the quest for a "theory of everything" that would explain all known physical phenomena in a single, coherent scheme.

We now know that electric current or, more fundamentally, *moving electric charge*, is the principle source of magnetism. There does not appear to be any basic, intrinsic property of matter that leads to magnetism in the way that electric charge leads to electrical phenomena. Rather, magnetism is simply one aspect of the phenomena associated with the existence of electric charge. And, again, the phenomena of magnetism arise not from electric charge alone but from *moving* charge.

You might be wondering what all this has to do with the familiar magnets you use to stick notes to your refrigerator. Does their magnetism involve electric charge? It sure does. The magnetism of everyday magnets results from the motion of electric charges, in this case the electrons moving about in the atoms that make up the magnet. Those motions make each atom a miniature magnet. That, in fact, is true of most atoms. But in a few materials, notably iron, the individual atomic magnets interact in such a way that their magnetic effects reinforce, resulting in the large-scale magnetism we see in ordinary permanent magnets. There's really not much difference between a permanent magnet and an electromagnet—one made by winding a coil of wire and passing electric current through it. In both cases the magnetism results from the motion of electric charges.

The relation between moving charge and magnetism is a two-way street. Moving charge not only creates magnetism but it also responds to magnetism. Fundamentally, the interaction of magnetism with matter results from a force that magnets exert on *moving* charge. Place an electron—a fundamental constituent of matter that carries negative charge—near a magnet and nothing happens. But if it's *moving*, the electron experiences a force that depends on the strength of the magnet and on the electron's motion. That's why your TV has circuitry to compensate for Earth's magnetism. The same terrestrial magnetism that guides compass needles would alter the paths of the electrons that beam through your TV tube to create the picture you see on the screen. Once again, magnetism is about *moving electric charge*.

Are there other relations between electricity and magnetism? In the 1820s the English physicist Michael Faraday pursued this question and by 1831 he had an answer. Because electric current produced magnetism, Faraday wondered if, conversely, magnetism might produce electric current. He made an electromagnet by winding a coil of wire and passing electric current through it. He placed another coil nearby and connected it to a meter that would register a flow of current in this second coil. Nothing happened. But then Faraday noticed a deflection of the meter when he interrupted the current in the first coil. That interruption represented a change in the current and hence in the magnetism of the first coil. Further experimentation convinced Faraday that he had discovered a new phenomenon, wherein *changing* magnetism can produce electric current. Today we call this phenomenon *electromagnetic induction*, and we exploit it in huge generators that spin wire coils in the presence of magnets to generate the electricity that powers our homes and industries. We also use it to read information from our videotapes and computer disks, whose changing magnetic patterns induce currents in a wire coil placed near the moving tape or spinning disk.

Implicit in the preceding paragraphs are four fundamental statements about electricity and magnetism. I'll state them very loosely here, but later we'll take a more careful and complete look at them:

- Electric charges exert forces on each other.
- Magnetic poles exert forces on each other, but there are no isolated magnetic poles.

- Moving electric charge gives rise to magnetism.
- Changing magnetism gives rise to electricity.

These four statements look rather different, but they actually comprise two pairs that reflect a profound symmetry between electric and magnetic phenomena. We need to develop that symmetry more fully in order to answer our question about the nature of light.

Fields

Suppose Earth suddenly vanished. How and when would the Moon, a quarter-million miles away, "know" that it should abandon its circular orbit—the result of Earth's gravity—and begin the straight-line motion that Newton's laws tell us it should take in the absence of a force? In Newton's view, the Moon would know about Earth's disappearance instantaneously, because in Newton's description of gravity Earth "reaches out" across empty space and immediately "pulls" on the Moon. Remove Earth, and that pull—the gravitational force—disappears instantly. This Newtonian idea is called *action at a distance* for obvious reasons.

There's another way to view the Earth–Moon interaction, a way that at first may seem needlessly complicating and abstract. But it will be crucial in developing your understanding of light and will become absolutely essential in the context of relativity. Even at this early stage, you may find this new view more philosophically satisfying.

Here's the issue: The action-at-a-distance description of gravity requires the Moon to know what's happening, right now, at the location of the distant Earth. But how can this be? It would be much more believable if the Moon only needed to know about its immediate vicinity and responded only to *local* conditions. Enter the concept of *field*. We imagine that Earth creates a kind of influence, called a *gravitational field*, everywhere in the space around it. Put an object in Earth's vicinity and it experiences not a mysterious action-at-a-distance pull from Earth but rather the gravitational field right at the point where the object is. It responds to that field by experiencing a force toward Earth's center, a force whose strength depends on the object's mass and on the strength of the gravitational field. Since gravity decreases with distance from Earth, so must the strength of the field. We can represent the gravitational field by drawing arrows that show its strength and direction at selected points (Figure 4.4).

Conceptually, in introducing the field concept we've replaced the direct but philosophically disturbing action-at-a-distance force of gravity with a force that arises *locally*, from the gravitational field at any point in space. Rather than Earth exerting forces directly, we have the more complex situation in which Earth creates a gravitational field in its vicinity and objects respond to that field. What we gain from this complexity is a new simplicity: now an object doesn't have to know about the situation at some distant place but only about what's happening in its immediate vicinity.

Of course, the ultimate outcome remains unchanged. The Moon, a spacecraft, and a falling apple still behave as Newton predicted. Only our description of how that occurs has changed; now these objects respond to Earth's gravitational field right where they are, rather than to Earth itself. But for now the action-at-a-distance and field perspectives predict exactly the same physical results.

Now back to the question What if Earth suddenly vanished? We can't answer that question for sure at this point, but at least the field

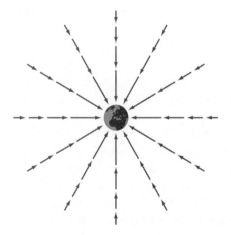

Fig. 4.4 A representation of Earth's gravitational field. The length of each arrow gives the strength of the gravitational field at its location, and the arrows point in the direction of the field, namely, toward Earth's center. The field exists everywhere, but I can only show arrows to represent it at a few selected points.

concept gives us some wiggle room. What happens at Earth's location isn't important to the Moon; it's the local gravitational field that determines the Moon's motion. So does the gravitational field everywhere vanish instantaneously when Earth does? Or does it have some sort of independent existence so it takes awhile for the field way out at the Moon's location to learn about Earth's demise? In fact, the answer is the latter, as we'll explore in later chapters. For now, though, suffice it to say that the field concept gives us a new and important physical entity—in this case the gravitational field— that can have, at least temporarily, an independent existence.

Electromagnetic Fields

I could have phrased the entire preceding section in terms of the electric force between charges rather than the gravitational force between Earth and Moon. I chose the latter only because you're more familiar with gravity. Having established the concept of the gravitational field, though, the *electric field* follows directly by analogy. We say that one electric charge creates an electric field everywhere in space and that a second charge responds to the field in its immediate vicinity. Again we get the same physical happenings as predicted by the action-at-a-distance electric force, but now the behavior of a charge is determined by *local* conditions—in this case the local electric field. Again we can speculate about what would happen if one charge suddenly vanished or just moved a bit; would a second, distant charge respond immediately or would it take a while for the local field to learn about this change and thus to adjust to it?

If electric charges create electric fields, surely magnets must create magnetic fields. And they do. Given our understanding of magnetism, though, we can put all this more carefully: Electric charges create electric fields, while *moving* electric charges create, in addition, magnetic fields. And we can rephrase Faraday's discovery as a statement about fields: the loose phrase "changing magnetism gives rise to electricity" becomes "A changing magnetic field creates an electric field." This last statement has an interesting implication: no longer is electric charge the sole source of electric phenomena, in particular of electric fields. Now something quite different, namely a changing magnetic field, can also create an electric field. So electric fields have *two* possible sources: electric charge and changing magnetic fields. Notice that the second of these entails a direct relation between the fields themselves, where one directly creates the other without the intermediary of electric charge or any other matter.

Let's look in more detail at the last two of our earlier statements about electromagnetism, now recast in terms of electric and magnetic fields:

- Moving electric charge creates a magnetic field.
- A changing magnetic field creates an electric field.

There's something vaguely unsatisfying, something unsymmetrical, about these two statements. If a changing magnetic field creates an electric field, might not a changing electric field create a magnetic field? That question occurred to the great Scottish physicist James Clerk Maxwell in the mid nineteenth century. Maxwell had good reason to suspect that a changing electric field would indeed create a magnetic field. First, our statements when expressed as mathematical equations look much more similar than the English versions, and the one virtually cries out for a mathematical term meaning "changing electric field" where the other has a term "changing magnetic field." Second, Maxwell saw that the equations were not fully consistent with the known fact that electric charge is conserved. He found that he could regain consistency by adding a term "changing electric field" in the statement/equation about the source of magnetic field. So he did, positing that his modified set of four statements provides a complete and consistent description of all electromagnetic phenomena. In Maxwell's honor, we now call those four statements collectively *Maxwell's equations*.

Electromagnetic Waves

So what's the big deal? Now a changing electric field creates a magnetic field and, as we knew from Faraday's work, a changing magnetic field creates an electric field. The big deal, as Maxwell soon deduced, is that this ability of each field to create the other results in a wondrous new phenomenon. Suppose, somehow, a changing electric field gets started. We might make such a field by grabbing an electron or other electrically charged object and moving it back and forth. That changing electric field creates a magnetic field. In general, that magnetic field will also be changing, so it creates an electric field. In general, that electric field will also be changing, so it creates a magnetic field. And so forth. You get the picture: once a changing field of either type appears, a self-perpetuating system of electric and magnetic fields is set into existence. Once they get started, the fields take on an independent existence, no longer associated with the electric charge or whatever it was that got them started.

In fact, Maxwell realized, this self-perpetuating structure of electric and magnetic fields wouldn't just sit there; rather, it would propagate through space as an *electromagnetic wave*. Like the other waves we've examined, such a wave would carry energy but not matter. Now Maxwell was not particularly interested in practical applications for these electromagnetic waves; he just wanted a complete, consistent description of electromagnetic phenomena. But in fact he had discovered something immensely useful. In 1887 the German Heinrich Hertz, again in the spirit of pure scientific inquiry, sought to confirm Maxwell's theory. He set up an electric circuit with a power source that made electrons wiggle back and forth, creating a changing electric field. A few feet away sat another circuit with no source of power. When the first circuit was energized, a spark jumped in the second circuit—indicating the arrival of energy via Maxwell's electromagnetic waves. Thus, Hertz vindicated Maxwell's theoretical hunches.

Was this work practical? Not yet. But in the 1890s the Italian-Irish physicist Guglielmo Marconi took up the challenge and used Maxwell's electromagnetic waves to establish so-called wireless telegraphy as a practical means of communication over long distances. In 1901 a triumphant Marconi sent electromagnetic signals from England to North America. Our modern radio, television, cell phones, satellite dishes, microwave ovens, electronic car keys, wireless computer mice, police radar, garage door openers, radio telescopes, and countless other technologies are all based on electromagnetic waves. All owe their existence to the wondrous interactions between electric and magnetic fields as described by Maxwell's equations.

My brief history of electromagnetic wave technology was just an aside to help you appreciate the immense practical significance of Maxwell's work. An even more philosophically profound result awaits us next.

Let There Be Light

Maxwell not only recognized the possibility of electromagnetic waves, but he also explored their properties as predicted by his equations. Among those properties is the speed of electromagnetic waves, which Maxwell could easily calculate from his equations. Those equations contain references to the electric and magnetic fields, of course, and also to electric charge and electric current. But they also contain two numbers that set the overall strength of the electric and magnetic forces. These two numbers represent fundamental properties of nature. Their values were determined experimentally, the electrical number from experiments on the electric force between charged objects and the magnetic number from experiments with electric circuits and the magnetism they produce. Those experiments, done in the eighteenth and nineteenth centuries using the simple apparatus of the time, were all that was needed to establish the values of the electric and magnetic numbers that appear in Maxwell's equations.

When Maxwell proceeded to calculate the speed of his electromagnetic waves, he found that it depended only on the values of these electric and magnetic numbers—quantities that, again, had been determined in simple laboratory experiments involving electric charge and electric current. When he worked out the value for the speed of his waves, Maxwell found it to be very nearly 300 million meters per second, or about 186,000 miles per second—and equal to the speed of light! That speed had been known since the late seventeenth century, and by Maxwell's time it was solidly established to an accuracy of a few percent. So there was no question: the speed of Maxwell's electromagnetic waves was the speed of light. Maxwell's conclusion was inescapable: *light itself is an electromagnetic wave!* In one brilliant stroke, Maxwell brought the whole science of optics under the umbrella of electromagnetism.

What a remarkable achievement is Maxwell's! First, he establishes a fully consistent theory of electromagnetism, expressed succinctly in the four Maxwell equations. He finds that his theory predicts an entirely new phenomenon—electromagnetic waves. Next he calculates the speed of those waves and recognizes that it's equal to the known speed of light. So now, some 60 years after Young's interference experiment established that light consists of waves, Maxwell has the answer to our question, Waves of what? Light consists of waves of electricity and magnetism, or electromagnetic waves.

How is light related to the electromagnetic waves of Hertz and Marconi and of modern communications? Simple: They're essentially the same thing. The only difference is the frequency with which the wave vibrates, or equivalently, the wavelength (the distance between wave crests, which decreases as the frequency increases). I've already mentioned that light waves have a wavelength about 20 millionths of an inch. AM radio, in contrast, has a wavelength of about 300 yards, while the waves used in FM radio have a wavelength of about 3 yards. The microwaves that cook your food have a wavelength of 3 inches, and police radar operates at a wavelength of about one-fifth of an inch. In fact, there's a whole continuous spectrum of electromagnetic waves, starting with the longest wavelength radio waves on down through the microwaves used for cooking, radar, and satellite communications. Shorter still and we're in the infrared—electromagnetic waves with wavelength slightly longer than those of visible light. You can feel infrared waves from a hot stove burner, clothes iron, woodstove, and other hot objects. Special infrared cameras can image "hot spots" indicating disease in the body or energy loss in a house. Still shorter and we're in the visible range of electromagnetic waves, from about 25 millionths of an inch for red light to about 14 millionths of an inch for deep violet. Beyond that is ultraviolet, which you can't see but which produces sunburn; then x-rays, and finally the penetrating gamma rays produced in nuclear reactions. We give all these ranges of electromagnetic waves different names out of convenience because of the different uses they have and the different means of producing them, but fundamentally they're all the same: all are electromagnetic waves, consisting of self-perpetuating structures of electric and magnetic fields, and all propagate at exactly the same speed. That speed, which we call the speed of light and denote by the symbol c, is really a universal speed of all electromagnetic waves. (That c comes most likely from the Latin *celeritas*, meaning swiftness.)

One caveat: c is the speed of electromagnetic waves *in a vacuum*. When electromagnetic waves travel through matter, they interact with electric charges in the matter (because light is an *electro*magnetic wave) and this slows down the wave. The phenomenon of refraction, for example, occurs because light moves more slowly in the plastic of your contact lenses or the glass of your camera lens than it does in air or vacuum. When I use the term "speed of light" in this book, I'll always mean the speed of light in vacuum unless I explicitly state otherwise. The speed of light is such a fundamental quantity that it's useful to know its numerical value. Again, that value is very nearly 300 million meters per second; 186,000 miles per second (more than seven times around the Earth in 1 second); or, in units appropriate to high-speed computers, about 1 foot per nanosecond (billionth of a second). It's also, in units useful to astronomers and in discussing relativity, exactly 1 light-year per year. More on this later.

Making Electromagnetic Waves

How do we make electromagnetic waves? How does nature do it? What we need to get a wave started is a changing field, either electric or magnetic. So in principle we could make an electromagnetic wave by taking an electrically charged object or a magnet and moving it back and forth, or spinning it, or whatever. The frequency of the wave would be the frequency of our back-and-forth or spinning motion and that would also determine its wavelength. (A detailed look at Maxwell's equations shows that the charge or magnet must be *accelerating*, that is, it must be in a state of *changing* motion. Uniform motion won't do, for reasons that relativity will make obvious.)

In practice, we make the relatively low-frequency electromagnetic waves in the radio region of the electromagnetic spectrum using electric circuits that feature rapidly alternating electric currents flowing in the antennas of radio transmitting systems. In your microwave oven and in radar units, electrons circle around billions of times a second in a special device called a magnetron, in the process emitting electromagnetic waves because circular motion is accelerated motion. The frequency of these waves is that of the electron's circular motion and corresponds to wavelengths of several inches and shorter. The still higher frequency, shorter wavelength infrared waves arise naturally from vibrating molecules. And most visible light, as well as ultraviolet, comes from individual atoms, as electrons jump about among different energy levels in the atom. In a medical x-ray unit, high-speed electrons slam into a metal target; their abrupt stopping constitutes the acceleration (change in motion) that produces electromagnetic waves in the x-ray region of the spectrum. Finally, energetic and extremely rapid processes within atomic nuclei produce the so-called gamma rays, electromagnetic waves with the highest known frequencies and shortest wavelengths. Again, the names we give these different electromagnetic waves are just for convenience; all the waves are basically the same, being self-perpetuating structures of electric and magnetic fields that propagate at the speed of light, *c*. Only the frequencies and corresponding wavelengths distinguish the different types of waves.

A Brief History of Physics

So here we are, in the late 1800s, with what looks like a complete understanding of all aspects of physical reality. That understanding springs from two different branches of physics: the mechanics (laws of motion) of Newton and the electromagnetism of Maxwell. (Of course many others were involved in establishing both branches, but for brevity I'll use only the names of Newton and Maxwell.) Many phenomena, from the behavior of gases to the motion of planets to the properties and propagation of sound, have Newtonian explanations. Electromagnetic phenomena, including all of optics (the study of light), follow from Maxwell's equations. Historically, Newton's mechanics came first, with the years from roughly 1600 to 1750 seeing the greatest advances in the Newtonian understanding of the Universe. Electromagnetism followed, with major advances from 1750 to nearly 1900. Together, these two distinct branches seemed to explain all of physics.

But do they really explain all of physics? And are their explanations consistent? Recall that Newtonian mechanics obeys a relativity principle—meaning that Newton's laws work equally well in any uniformly moving frame of reference. Is the same true of Maxwell's laws of electromagnetism? You now have enough background that these and related questions can lead you straight to an understanding of Einstein's thought.

CHAPTER 5

ETHER DREAMS

• • •

You've just seen how Maxwell's electromagnetism leads to electromagnetic waves, including light, and predicts that these waves should move with speed *c*, about 186,000 miles per second. But 186,000 miles per second relative to what? In the preceding chapter we asked the analogous question for sound, Seven hundred miles per hour relative to what?, and had an obvious answer. Sound moves at 700 miles per hour relative to air, the medium in which sound waves are a disturbance. So what's the corresponding answer for light? (When I say "light" in this context, I mean any electromagnetic wave, including visible light. I'll often use "light" and "electromagnetic wave" interchangeably.)

Before we try to answer this question, let me emphasize why it's so important. Without an answer, Maxwell's equations are on shaky ground because their prediction of electromagnetic waves moving with speed *c* is meaningless. If we can't say with respect to what that speed *c* is measured, then the statement "Light goes at speed *c*" is pretty vacuous. We're now going to explore possible answers to the question, Speed *c* relative to what? If you find you don't like those answers, then I challenge you to come up with your own! Keep in mind that we need an answer if Maxwell's electromagnetic theory—a theory that is supposed to explain a great many important physical phenomena including much of modern technology—is to have a solid grounding.

Enter the Ether

Nineteenth-century physicists, heady with the successes of Newtonian mechanics, naturally assumed that electromagnetic waves are like sound and all other known waves, namely, that they are disturbances of some medium. They called that medium the *ether*. Then, Speed *c* relative to what? had an obvious answer: Light goes at speed *c* relative to the ether. An immediate corollary is that if you're moving through the ether, then light's speed relative to you won't be *c*—just as the speed of sound waves measured by someone moving through the air is different from the 700-miles-per-hour-speed of sound relative to air.

Let's take a closer look at this ether. Ether is supposed to be to electromagnetic waves what air is to sound waves—the medium in which the waves are a disturbance. Now sound waves are possible only where there's air, which is why astronauts on the airless moon can't communicate by talking. But light reaches us from the far corners of the Universe, so ether must be everywhere. Ether must be a kind of tenuous, transparent "background" that pervades the entire Universe. It fills the space between the planets, stars, and galaxies. Because light propagates through transparent substances like air and water, ether must permeate the tiny spaces between and even within atoms.

This ubiquitous ether would need some unusual properties. It would have to be a fluid like air or water, rather than a solid, since things move through it. Further, it should be a very tenuous fluid, because it offers no resistance to the motion of planets. If it did, Earth and the other planets would lose energy and eventually spiral into the Sun. But at the same time ether would need to be very stiff, because the speed of light is so big. You can see this if you imagine a stretched spring, Slinky, or rubber band. Send a wave pulse down the spring by disturbing it briefly and then letting go. The more you stretch the spring, the faster the pulse travels. An analogous situation holds for sound waves in different media; sound travels faster in water than in air, for instance, because water is "stiffer" in the sense of the spring with more stretch. And an analogy holds for electromagnetic waves in the ether. For those waves to have the high speed of 186,000 miles per second, the ether must be very stiff indeed. The two ether properties of being tenuous, to minimize resistance, and being stiff, to maximize wave speed, aren't easily reconciled. Ether would have to be a rather unusual material.

Don't Like the Ether? Try Another Answer

Its contradictory properties make ether an improbable substance, but so ingrained was the mechanical paradigm and so crucial is it to answer the question, Speed c relative to what? that nineteenth-century physicists saw no alternative to the ether. You may think you see an alternative, namely, do away with the ether entirely and let electromagnetic waves propagate through empty space. If you go down that road, then I'm going to ask you for an answer to the question Speed c relative to what?, and I'm going to expect a good answer that's consistent with physical experiments.

Here's a seemingly good answer: Light goes at speed c relative to its source. Eliminate the ether, that is, and let light propagate through empty space at 186,000 miles per second relative to whatever object emitted the light. An observer at rest with respect to that light source would measure c for the light's speed, but an observer for whom the source was moving would obviously measure a speed different from c. This idea is pretty simple. An analogy would be a baseball pitcher who can throw a 100-mile-per-hour fastball. Thrown from the pitcher's mound, the ball moves at 100 miles per hour relative to the pitcher. Since the pitcher is standing on the ground, that amounts to 100 miles per hour relative to Earth. But put the pitcher in an open car going 50 miles per hour. Again, the ball goes 100 miles per hour relative to its source, namely the pitcher. But relative to observers on the ground, it would be going 150 miles per hour. (Or so it seems; more on this later!)

So does the measured speed of light depend on the speed of its source relative to the observer? No, it doesn't. A host of experiments and observations confirm this. A particularly straightforward confirmation comes from astronomical observations of double-star systems. Unlike our Sun, about half the stars in the Milky Way galaxy are locked in a gravitational embrace with another star. These double stars orbit around each other with orbital periods ranging from hours to years. In some cases the system lies with the plane of its stars' orbits in line with our view from Earth. Then as the stars orbit, one moves toward us while the other moves away; half a period later this situation is reversed (Figure 5.1). We can analyze the stars' motions in detail by studying slight shifts in the wavelength (i.e., color) of light as we observe it. These changes arise because of the star's motions relative to Earth.

Now even the nearest stars are very far away—so far that light from them takes many years to reach us. A slight difference in the speed of light from the two stars in a double-star system would show up as a long time delay between our receiving light that was, in fact, emitted simultaneously from the two stars. We would have to compensate for this difference in light travel times to infer the stars' motions. But we need no such compensation! The stars move exactly as Newton's law of gravity says they should, based on the light as we receive it. There's no increase in the speed of light from the star that's momentarily moving toward us and no decrease for the star that's moving away from us. The speed of light simply does not depend on the motion of its source.

So what's wrong with our baseball analogy? What's wrong is that light doesn't consist of particles like miniature baseballs. Rather, as you know, it consists of waves. Think about sound waves for a

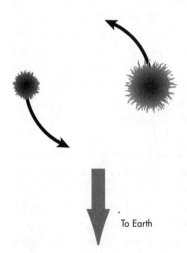

Fig. 5.1 A double-star system, showing the two stars orbiting each other. At the moment pictured, the left-hand star is approaching Earth and the right-hand star is receding. If the speed of light depended on the speed of its source, light from the left-hand star would reach Earth sooner.

To Earth

minute. The speed of sound waves is 700 miles per hour relative to
the medium, namely air, in which sound is a disturbance. If a source
of sound—say a fire truck's siren—moves through the air at 60 miles
per hour, the speed of the sound is still 700 miles per hour relative
to the air. It doesn't pick up the additional 60 miles per hour of the
fire truck. Apparently light is like that. It must move at c relative to
the medium, namely ether, in which it is a disturbance. If a source
of light moved through the ether, the speed of the light would nevertheless still be c relative to the ether.

(By the way, something about the sound from a moving source
does change, namely, its pitch. When the fire truck is coming toward
you, the siren sounds like it has a higher pitch. When it moves away,
the pitch you hear is lower. The analogous shift for light is what lets
us measure motions in double-star systems.)

In disabusing you of the notion that the speed of light might be
c relative to its source alone, I reverted naturally to the language of
ether—the medium in which light waves are purportedly a disturbance. Despite the problematic aspects of the ether, what else are
we (or, more precisely, nineteenth-century physicists) to do? Light,
after all, consists of waves, and all other known waves are disturbances of some medium. The speed of each type of wave—sound
waves, water waves, stadium waves, earthquake waves, etc.—is its
speed relative to its particular medium. Why not the same for light?

If you're still troubled by the all-pervasive, tenuous, light-wave-
supporting ether, then again I challenge you to tell me relative to
what light travels at speed c. That's a harder challenge now you
know that the answer cannot be "relative to its source." So for now
we'll continue to follow the nineteenth-century physicists' line of
thought, picturing light waves as disturbances of an ether that pervades the entire Universe. If you really don't like the ether, though,
hold on—eventually you'll be vindicated!

A Broader Question

We were led to the ether concept by questioning relative to what
light travels at speed c. I want to convince you now that this ques-

tion is really a special case of a more general one: In what frame of reference are the laws of electromagnetism (i.e., Maxwell's equations) valid? The two questions are related because one prediction of the laws of electromagnetism is that there should be electromagnetic waves and that they should go at the speed of light, c. When we answer "ether" to the question "Relative to what does light go at speed c?", we're saying that Maxwell's prediction of electromagnetic waves that go at c is really valid only in a frame of reference at rest with respect to the ether. Observers in a reference frame moving through the ether will measure some other speed for light relative to themselves; thus for them, in the context of their reference frame, the predictions of Maxwell's equations won't be valid. That's why my two questions, one about the speed of light and the other about the validity of Maxwell's equations, are essentially equivalent. Maxwell's equations predict electromagnetic waves going at c, so those equations can only be valid in a frame of reference where one will, in fact, measure c for the speed of light.

So what's the answer to our new question, In what frame of reference are Maxwell's equations valid? It's obvious: In the context of nineteenth-century physics, there's only one frame of reference in which Maxwell's prediction about electromagnetic waves is valid, and that's a frame of reference at rest with respect to the ether.

Dichotomy in Physics

In Chapter 3 we discovered the principle of Galilean relativity, which states that Newton's laws of motion are valid in all uniformly moving frames of reference. So if we ask explicitly in what frame of reference Newton's laws of motion are valid, then the answer is "in any uniformly moving frame of reference."

This question we've just asked about Newton's laws of motion is exactly the same question we asked in the previous section about Maxwell's laws of electromagnetism. But there we found quite a different answer. Maxwell's electromagnetism, it seems, isn't valid in just any frame of reference. Rather, the laws of electromagnetism should be valid only in one very special frame of reference—a frame of reference at rest with respect to the ether. Put another way, the laws of motion obey a relativity principle but the laws of electromagnetism seem not to. So although there's no experiment we can do with the laws of motion to answer the question, Am I moving?, there should be electromagnetic experiments that can answer this question. That is, the concept of absolute motion is meaningless for mechanics, but apparently it has meaning for electromagnetism. Although there's no privileged state of motion for mechanics, there seems to be a privileged state for electromagnetism—namely, being at rest relative to the ether.

Why the dichotomy? Why should one main branch of physics (mechanics) not care about states of motion, while the other (electromagnetism) does? Wouldn't it be simpler and more coherent if both branches of physics obeyed the relativity principle, or both didn't?

Think back to the second chapter, where I asked about a tennis match played on a cruise ship, on Venus, and on a planet in a distant galaxy moving away from Earth at nearly the speed of light. You wisely and logically agreed that it made perfect sense to expect that tennis playing—a manifestation of the laws of motion—would work the same in all those contexts or, as we would now say, in all those different reference frames. That is, you intuitively accepted the principle of relativity as applied to the laws of motion. Then I asked about heating a cup of tea in a microwave oven, and you agreed that the microwave oven, like the tennis ball, should also behave the same way in the different reference frames. But it's electromagnetism, not mechanics, that governs the microwave oven. Through our exploration of the question, Relative to what does light go at speed c?, we've just found that the laws of electromagnetism seem not to obey the relativity principle. That is, the laws of electromagnetism should not be valid in all reference frames—and electromagnetic experiments should therefore give different results in reference frames that are in different states of motion. So the microwave oven shouldn't work the same way in that distant galaxy moving at nearly c as it does on Earth!

What's wrong with your intuition from Chapter 2? Again, the difficulty is with the ether. When you blithely agreed that the microwave oven should work the same way everywhere, you weren't taking the nineteenth-century view, with its unique reference frame of the ether, in which frame alone the laws governing the oven should be valid.

So why not get rid of the ether, now troublesome not only because of its improbable characteristics but because it's also the cause of an illogical and unsatisfying dichotomy between two branches of physics? That dichotomy runs counter to your good sense that the laws governing physical reality, whether tennis balls or microwave ovens, should work the same everywhere and regardless of one's state of motion. If you abandon the ether, then you could eliminate that dichotomy.

But if you abandon the ether, then I challenge you once again to answer the question, Relative to what does light go at speed c? or, equivalently, In what frame of reference are the laws of electromagnetism valid? If you give the same answer that I'll happily accept for the laws of motion—"in all uniformly moving reference frames"— then you'll vindicate your intuitive sense from Chapter 2 that the microwave oven should work the same in all states of motion. But if you give that answer, you'll find yourself on the edge of a philosophical abyss. That's because you'll be insisting on a seeming contradiction: that two different observers must each find valid the Maxwellian prediction that light waves move at speed c—*even if those observers are moving relative to each other*! Better not go there, at least not yet; instead, we'll stick for now with the nineteenth-century ether concept and explore further its implications.

Earth and Ether

Following nineteenth-century physicists, we've established ether as the medium through which light waves propagate with speed c, and we recognize therefore that Maxwell's equations of electromagnetism can only be valid in a frame of reference at rest with respect to the ether. So a logical question arises: How is our Earth moving relative to the ether?

It's pretty obvious that we aren't moving very fast through the ether, because if we were then we would see obvious differences in the speed of light coming from different directions. For example, if Earth were moving at 90 percent of c, then light from stars in the direction toward which Earth is moving would be going at $1.9c$ relative to us ($c + 0.9c = 1.9c$). But light from stars in the opposite direction would be going at only one-tenth of c ($c - 0.9c = 0.1c$). That difference would be patently obvious. But if Earth were moving much more slowly through the ether—at a tiny fraction of c—then we might not notice the difference unless we looked very carefully. So again the question: How is Earth moving relative to the ether? That is, how fast and in what direction is Earth's motion relative to the ether?

We begin with an even simpler question: Is Earth moving relative to the ether? That one has a yes-or-no answer. Either Earth is moving relative to the ether, or it isn't.

Insulting Copernicus

Consider first the possibility that Earth isn't moving relative to the ether. I can think of only two ways for this to be the case. First, the ether might be a fixed substance that extends throughout the Universe. Then Earth alone among all the cosmos would be at rest relative to the ether. I say "alone" because all other celestial objects—the Moon, Mars, Venus, the other planets, the Sun, other stars in our galaxy, and the other galaxies in the Universe—all are moving relative to Earth. So if Earth is at rest relative to the ether, then it alone is at rest. That makes us pretty special. If we're the only beings at rest relative to the ether, then Maxwell's equations are valid only for us, and only we measure c for the speed of light. Observers on other celestial bodies measure different speeds for light in different directions, and for observers moving very fast relative to Earth—like those in distant galaxies—that effect must be dramatically obvious.

Copernicus would turn in his grave! It's hard to imagine a worse insult to the Copernican revolution than to make our planet so special that one of the two main branches of physics is valid only on Earth. I spent most of Chapter 3 presenting a history of science that led steadily away from the notion of Earth being a special, privileged place in the Universe. Do you really want to return to parochial, pre-Copernican ideas? Do you really think you and your planet are so special that, in all the rich vastness of the Universe, you alone can claim to be "at rest"?

On purely philosophical grounds, we should reject the notion that Earth alone could be at rest relative to the ether. Now, philosophy isn't science, and I hasten to add that there's plenty of good scientific evidence to support this view. For example, we observe light-emitting processes in distant stars and galaxies that seem to work the same there as they do here on Earth. That suggests we don't have any special status vis-à-vis the laws of electromagnetism. So we can confidently reject the idea that Earth alone is at rest relative to the ether.

Ether Drag

It might still be possible for Earth to be at rest relative to the ether if our planet somehow "dragged" the surrounding ether with it. Presumably other planets and celestial bodies would do the same, so each would be at rest relative to its local blob of ether. Then observers everywhere and in different states of motion would find the laws of electromagnetism to be valid, and no one would have any claim to be special. Copernicus would be a lot happier with that!

So does Earth drag the ether with it? Astronomical observations dating to 1725 provide a clear answer. A simple analogy will help you understand these observations. In Figure 5.2a you're standing, holding an umbrella in the rain. Obviously the best approach to keeping dry is to hold the umbrella directly overhead. But what if you run through the rain? Now it's better to tilt the umbrella, holding it at an angle (Figure 5.2b). Figure 5.2c shows why: viewed from your frame of reference, the rain is coming down at an angle, and you want to hold the umbrella so the rain still hits the umbrella top straight on. Since the rain is falling at an angle, you should hold the umbrella at the same angle. If you run in the opposite direction, you should still hold the umbrella tilted at the same angle in front of you, but now this will be a different absolute direction.

However, suppose that somehow you drag a big blob of air with you as you run—so big a blob that rain falling into it has time for

(a) Stand still   (b) Run

(c) In runner's frame   (d) With air drag

Fig. 5.2 An analogy for the aberration of starlight. (a) Standing still in vertically falling rain, you hold your umbrella straight overhead to keep driest. (b) Running, you tilt your umbrella. (c) The situation in (b), shown from the runner's frame of reference. In this frame, the rain falls at an angle. (d) If the runner drags a large blob of air, then rain entering the blob will take on the blob's motion, and thus will fall vertically relative to the runner.

the force of the moving air to accelerate it to your running speed before it hits you or your umbrella. Figure 5.2d shows the situation from your point of view. As you run through the rain, the rain outside your blob of dragged air falls at an angle as seen from your reference frame. But inside the blob, the moving air accelerates the rain until it shares the blob's motion. So now, relative to you, it's falling straight. The upshot is that you don't have to tilt your umbrella. Rather, you'll stay driest if you hold it right overhead.

You can see from the foregoing discussion that an umbrella in a rainstorm is a useful device for determining whether you drag a big blob of air when you run—as if you don't already know the answer. But let's suppose you don't. So you set out in a rainstorm to answer the question. You run through the rain and hold your umbrella first over your head, then at an angle. You find out which approach keeps you driest. If the overhead approach is best, then, as in Figure 5.2d, you can conclude that you do drag a big blob of air with you. But if tilting the umbrella works best, then from Figure 5.2c it's obvious that you don't drag the air with you.

Running in the rain with an umbrella provides an analogy for astronomical observations that answer the question, Does Earth drag a blob of ether with it? The runner moving through the air is like Earth moving through the ether. The rain is like light from a distant star "falling" on Earth. The umbrella is an astronomer's telescope. And the question is, Do we need to point the telescope at an angle to compensate for Earth's motion? Actually, the question is slightly more subtle, since we can't stop Earth and compare the telescope angles with Earth stopped or moving. But what we can do is compare the angles for different directions of Earth's motion. Relative to the Sun, Earth is right now moving in some direction. Six months later, halfway around its orbit, the planet will be going in the opposite direction. If Earth doesn't drag the ether, then we'll need to change the angle of our telescope to observe the same star at six-month intervals, just as running in different directions in Figure 5.2c requires changing the direction of the umbrella's shaft. Such a change in the apparent direction to a star is known as *aberration of starlight*. If, on the other hand, Earth does drag the ether, then we keep the telescope in the same direction for both observations. In this case there's no aberration. So what happens? In fact, the telescope angle changes—a small change, to be sure, but readily detectable with astronomical measurements as early as 1725. We conclude that Earth does not drag ether with it.

It seems we've answered the question of whether Earth is moving relative to the ether. On philosophical as well as observational grounds we first ruled out the possibility that Earth alone is at rest relative to the ether. Now aberration of starlight shows that Earth doesn't drag a blob of ether with it. Those were the only ways we could conceive of Earth's being at rest relative to the ether. So we have to conclude that Earth *is* moving relative to the ether.

Given that fact, we can now ask how fast and in what direction Earth is moving. That's the question nineteenth-century physicists set out to answer. With a variety of clever experiments, they sought to measure Earth's motion through the ether. These experiments typically involved the propagation of light through water and had the added benefit of verifying the notion that moving substances— for example, flowing water—do not communicate their motion to the surrounding ether. That is, the experiments confirmed the absence of ether drag. But, curiously, they failed to detect Earth's motion through the ether. This disappointing result was attributed to the effects of the water on light, effects that apparently just cancelled out the sought-after indication of motion through the ether. The question of Earth's motion remained ambiguous.

Wrap-up: Physics at 1880

It's now about the year 1880, and here's how things stand. Together, Newton's mechanics and Maxwell's electromagnetism seem to explain all known physical phenomena. There's a philosophically disturbing dichotomy, in that a relativity principle holds in mechanics but not in electromagnetism, but that doesn't diminish the explanatory power of these two great branches of physics. Light is understood as an electromagnetic wave, propagating with speed c through a Universe-permeating medium called ether. Astronomical observations show that the speed of light does not depend on the motion of its source and that Earth must be in motion relative to the ether. The only thing remaining to solidify the picture of electromagnetic waves in the ether is to measure Earth's motion. But as of 1880 no experiment has succeeded in doing so. Then again, the experiments are difficult, they require great sensitivity, and there are complicating effects that may obscure the desired result. What's needed is a conceptually simple experiment that's sensitive enough to measure unambiguously Earth's motion through the ether. Such an experiment should, once and for all, lay to rest any nagging doubts about the ether.

CHAPTER 6

CRISIS IN PHYSICS

• • •

How are we to determine Earth's motion relative to the ether? If light moves with speed c through the ether then, as I've stressed before, observers who are moving relative to the ether should measure values other than c for the speed of light relative to themselves. So the most obvious way to detect Earth's motion through the ether is to measure the speed of light and see if it agrees with the value c predicted from Maxwell's equations. If it doesn't, then the observer must be moving relative to the ether, and the difference between the measured speed of light and c should reveal the speed of that motion. Unfortunately, no measure of light's speed in the nineteenth century was remotely precise enough to distinguish between the actual value c and a slightly different value caused by a modest earthly motion through the ether.

However, a moving observer will find not only that the speed of light is not quite c but also that light's speed depends on the direction of the observer's motion in relation to that of the light. As Figure 6.1 shows, an observer moving toward a source of light will measure a higher speed than c, while an observer moving away will measure a lower speed. So a more subtle approach to the problem is to try to detect *differences* in the measured speed of light in different directions. This approach avoids having to know the actual value with extreme precision; all that's important is measuring directly the difference in two values, without measuring the values themselves.

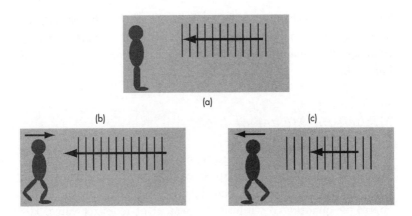

Fig. 6.1 Observers moving relative to the ether (gray rectangles) would measure different speeds for light. Vertical lines are crests of a light wave, and the length of the arrow represents the light's speed relative to the observer. (a) A stationary observer would measure c, the normal speed of light. (b) An observer moving toward the source of the light would measure a higher speed, and (c) an observer moving away, a lower speed.

Can we get any idea of how much difference to expect? Not entirely, because we don't know Earth's speed relative to the ether. That, after all, is what we're trying to find out. But we know that Earth orbits the Sun at some 20 miles per second, and because the orbit is circular, the direction of that motion changes throughout the year. So even if the Sun, by some bizarre coincidence, happened to be at rest in the ether, Earth would be moving at some 20 miles per second. And whatever Earth's speed, physicists knew it had to change by some tens of miles per second throughout the year, as Earth first heads in one direction at 20 miles per second relative to the Sun and then, 6 months later, at 20 miles per second in the opposite direction. If they could build a device to detect a speed of that magnitude, they could answer the question of Earth's motion through the ether.

Now, 20 miles per second sounds fast but it's slow compared with the speed of light, some 186,000 miles per second. Worse, it turns out that a successful measurement of Earth's motion requires detecting the *square* of Earth's speed in relation to the *square* of the

186,000-mile-per-second speed of light. That's like measuring the difference between the numbers 1 and 1.00000001. So great was the challenge that many nineteenth-century experimenters thought it impossible.

Michelson, Master of Light

Enter the Prussian-born American physicist Albert A. Michelson. Michelson's passion was the speed of light, and by 1879, while a young instructor at the U.S. Naval Academy, he had measured that important quantity to within 0.05 percent of its exact value. His expertise at precision measurement gave Michelson hope that he might be the first to measure unambiguously Earth's motion relative to the ether. Michelson transformed that hope into series of experiments, culminating in an 1887 version that remains among the most famous experiments in all of science. I'm going to spend some time discussing this experiment because it provides some of the most convincing evidence in support of Einstein's relativity. (For you, and for most physicists, that is. Whether the experiment influenced Einstein himself remains a matter of debate among historians of science. More on this later.)

Michelson was a brilliant and meticulous experimenter, and he invented a device that, to this day, forms the heart of instruments providing precise measurements of distance, time, and other quantities. This device, the *Michelson interferometer*, uses the interference of light waves to detect minute differences in the time light takes to travel two different paths. Michelson's interferometer can easily measure time differences less than the oscillation period of light— about one-thousandth of a trillionth of a second. With that kind of precision, Michelson knew he could detect variations in the speed of light due to Earth's orbital motion.

Conceptually, Michelson's experiment is simple. Two beams of light travel equal-length paths oriented at right angles to each other. Because of Earth's motion through the ether, the apparatus experiences an "ether wind," just as you feel a wind when you stick your hand out the window of a moving car, even on a calm day. Since light travels at speed c through the ether, this ether wind will affect the time it takes light to travel through Michelson's apparatus. Because the two beams travel paths that are oriented differently, the wind's effect on the two beams will, in general, be different. The idea is to detect and measure that difference.

I'll consider in detail only the simplest case, in which one light path lies along the direction of the ether wind (i.e., along the direction of Earth's motion through the ether) and the other lies at right angles. Now, for reasons that will soon be obvious, Michelson's light beams travel on round-trip paths. So the beam moving along the direction of the ether wind has to move both upstream, against the wind, and then downstream, with the wind. Going upstream, the light is slowed as measured by an observer on Earth, just as a boat traveling up a flowing river moves more slowly relative to the bank than if the river weren't flowing. Going downstream, with the ether wind, the light's speed will increase. So will the boat's. Now, you might think these two effects cancel out—but they don't. Here's why: Because it's moving slowly going upstream, the light (or the boat) spends more time being slowed by the ether wind (or the river current) than it does being sped up on the return trip. In fact, if the ether wind were blowing at the speed of light c, the light would never get anywhere and its round-trip would take forever! Similarly, if you rowed a boat upstream at just the speed of the current, you wouldn't get anywhere in relation to the riverbank. So the light's upstream/downstream trip will always take longer than it would in the absence of an ether wind.

Meanwhile, the other light beam is taking a round-trip journey at right angles to the wind. You might think the wind would have no effect on this beam, but that's not so. Again imagine a boat, which you're now trying to row to a point straight across a river. If you aim straight across, you won't get where you want to be, because the current will drag you downstream. So you have to aim a bit upstream and that makes the trip take a little longer. You have to do the same on your return trip, so in this case, too, the current slows the boat. But because on this perpendicular trip the current isn't affecting you as directly, the slowing-down effect is less than in the upstream/downstream trip. (A little work with ninth-grade algebra and the Pythagorean theorem could convince you rigorously of this conclusion.) So in Michelson's apparatus the light whose path lies along the ether wind should take longer to make its round-trip than the light whose path is perpendicular to the wind. That time difference is what Michelson set out to measure.

Even with Earth's 20-mile-per-second orbital speed, the time to travel Michelson's two paths is tiny—measured in billionths of a second. And the expected time difference would be about a million times smaller, so much so that even if one could measure the times, one couldn't possibly do so with enough accuracy to determine their difference. Here's where Michelson's genius comes in. Instead of trying to measure the two times, he went directly for the time difference—which, after all, is the significant manifestation of the ether wind. It was the nature of light itself—the *wave* nature of light— that let Michelson measure that time difference. Because light consists of waves, it exhibits interference—the phenomenon that, back in 1801, had convinced Thomas Young of light's wave nature. In describing Young's experiment in Chapter 4, I distinguished between two types of interference, constructive and destructive. These led, respectively, to the bright and dark bands in the pattern of light falling on Young's screen. Now here's the crucial point: constructive interference (bright light; wave crests in step) turns into destructive interference (dark; crests meeting troughs) if one of the two interfering light waves gets delayed just enough that its troughs fall back to where its crests had previously been. Because light's wavelength (distance between crests) is so small, it doesn't require much of a time delay to turn constructive interference to destructive. How much? About a thousandth of a trillionth of a second!

So here's Michelson's idea, as shown in Figure 6.2. Produce a single beam of light of a pure color (meaning a single wavelength). Split the beam in two, and send the resulting beams on two equal paths at right angles. How? With a half-silvered mirror, also called a *beam splitter*. This is just a lousy mirror that didn't get enough reflective coating. So only about half the light that hits it gets reflected, while the other half goes straight through. Michelson placed his beam splitter at a 45-degree angle to the original light beam. The result, shown in Figure 6.2, is to send half the light straight on and to

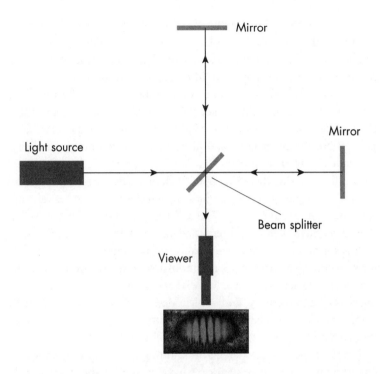

Fig. 6.2 In the Michelson–Morley experiment, a light beam splits and travels two perpendicular paths. Light returning from the mirrors is rejoined at the beam splitter, resulting in interference seen in the viewer. Photo below the viewer is the interference pattern from an actual Michelson–Morley setup.

reflect half of it perpendicular to its initial direction. Both light beams then travel on equal-length paths, bounce off ordinary mirrors, and return to the half-silvered mirror. Again, half of each beam is reflected and half goes through. The important point, as Figure 6.2 shows, is that portions of each beam come together again, traveling toward the viewer at the bottom of the picture.

If the beam paths in Figure 6.2 were exactly equal, and if there were no ether wind, then the beams would return exactly in step and the light waves would interfere constructively. An observer looking into the viewer would see bright light. But suppose the ether wind delayed one beam by exactly enough to make its troughs line up with the other beam's crests. Then we would have destructive interference and darkness in the viewer. That, in principle, is how Michelson's interferometer could detect Earth's motion.

Actually, it's both a bit more complex but also simpler than that. First of all, it's impossible to get the two ordinary mirrors at exactly right angles, and in any event the light rays in the beams aren't exactly parallel but diverge slightly and hit the half-silvered mirror at slightly different points. The result is that light takes a whole lot of different paths, differing slightly in length, on each of the two legs of Michelson's apparatus. Some of it returns in step, interfering constructively, and some of it returns out of step. The result is not simply bright or dark in the viewer but a pattern of alternating light and dark bands. (Figure 6.2 includes a photo of interference bands from a modern-day Michelson interferometer). Furthermore, it's impossible to get the distances to the two mirrors exactly the same. But all that does is to alter just which light interferes constructively and which destructively. Even if the path lengths aren't the same, we'll still get a pattern of light and dark bands. It's just that the positions of the bands will be a bit different.

None of these subtleties matters, though; in fact, they make the experiment easier, since we don't have to worry about getting the path lengths equal or the mirrors exactly perpendicular. The reason they don't matter is that we aren't really interested in the interference pattern itself, which includes the effects not only of the ether wind but also of the various imperfections in the instrument like path lengths and mirror alignment. But now suppose we rotate the whole apparatus through 90 degrees. The path that was initially along the ether wind is now at right angles to it, and the one that was originally at right angles is now along the wind. Whatever the relative timing for the light beams on the two paths was originally, the important point is that the timing should now *change*. That change is due entirely to the ether wind because nothing about the apparatus itself has changed except its orientation in the wind. And how do we detect that change in the relative travel times on the two paths? Simple: we watch the interference pattern. A change in travel times should result in a shift in the positions of the bright and dark interference bands. That shift, viewed as the apparatus is rotated, gives a direct indication of Earth's motion through the ether. The magnitude of the shift is a measure Earth's speed.

The Michelson–Morley Experiment

By 1880 Michelson, then on a 2-year study trip to Europe, had built his first interferometer. With this device, he knew, Earth's orbital motion should produce a shift of only a few percent of the distance between bands—barely detectable. Returning to the United States, Michelson secured a position at Case School of Applied Science in Cleveland. He soon began collaboration with Edward Morley, a chemist from nearby Western Reserve University (much later merged with Case to form Case Western Reserve University). By 1887—after enduring Michelson's brief but forced commitment to a "nerve specialist" for supposed "softening of the brain" and a disastrous fire at the Case School—Michelson and Morley had ready a much improved version of Michelson's interferometer. This device was mounted on a 5-foot square stone slab floating in liquid mercury, allowing easy rotation of the entire apparatus without disturbing the delicate light paths. Use of multiple mirrors greatly increased the effective lengths of the two light paths. As a result the Michelson–Morley experiment of 1887 was sensitive enough that a shift of nearly half the distance between light and dark fringes should occur as a result of Earth's orbital motion—a shift that would be obvious to the observer, as shown in Figure 6.3.

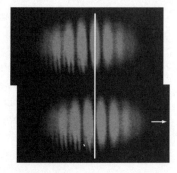

Fig. 6.3 Michelson and Morley should have seen a shift of nearly half a fringe, putting a dark band where a light band had been. Photos, from an actual Michelson–Morley setup, are offset to show the fringe shift.

Michelson and Morley performed their experiment by slowly rotating the apparatus as one of them walked around with his eye to the viewer. They repeated the experiment at different times so their entire lab would be oriented differently relative to the ether wind. And what was the result? A dismal failure: they never saw any significant shift in the interference bands!

Contradiction!

The starkly negative outcome of the Michelson–Morley experiment stands as one of the most important experimental results in all of science. To see why, remember where we are, logically, in 1887. The realization that light is a wave, specifically an electromagnetic wave propagating at speed c, raises the question of the medium in which light propagates. Nineteenth-century physics answers that question by proposing the ether as the medium for light and other electromagnetic waves. It then makes sense to ask about Earth's motion relative to the ether. Aberration of starlight shows that Earth can't be at rest relative to the ether, so Earth must be moving. Earlier experiments fail to detect that motion, but they suffer either from conceptual flaws or insufficient sensitivity. Now, in 1887, come Michelson and Morley with an experiment much more sensitive than what's needed to detect something that must exist, namely, Earth's motion through the ether. And yet the experiment fails. Earth *must* be moving through the ether, yet the Michelson–Morley experiment shows that *it isn't*. That's a pretty stark contradiction, and it shook the foundations of physics in the late nineteenth century.

I repeat: it's the foundations of physics—the basis of our whole understanding of physical reality—that are shaking, not some minor inconsequential details. Why is this contradiction so profound, so dire? Because it concerns a fundamental and sweeping prediction of one of the two basic branches of physics, specifically Maxwell's electromagnetism with its prediction of electromagnetic waves propagating at speed c. That prediction immediately gives voice to the question, Speed c relative to what? It's in attempting to answer that question that the contradiction inherent in Michelson–Morley arises.

If we can't resolve that contradiction, then there's something drastically wrong with our supposed understanding of physical reality.

No wonder nineteenth-century physicists sought at all costs to explain away the negative Michelson–Morley result. Not to do so would be to admit a logical fault so deeply ingrained as to threaten the entire edifice of physics. The experiment itself seemed beyond reproach, so physicists sought some explanation, some excuse, for its failure to detect Earth's motion. Michelson himself concluded disappointedly that Earth must be at rest relative to the ether after all, despite the apparently opposite implication of the starlight aberration observations. Others made more radical suggestions. In particular, the Dutch physicist H. A. Lorentz and the Irish physicist George Fitzgerald independently proposed that the ether squeezes objects moving through it, contracting them in the direction parallel to their motion. This contraction would shorten the ether-wind-aligned path in the Michelson–Morley experiment and thus reduce the travel time for light along that path. If the contraction were just right, the effect would eliminate the time differences on the two paths and would therefore explain the negative result of the experiment.

The Lorentz–Fitzgerald contraction would be very small, amounting to a decrease of only about 3 inches in Earth's 8,000-mile diameter. But that's all it would take to explain the Michelson–Morley result and to restore the logical consistency of physics in the nineteenth century. But given ether's tenuous nature, why should the contraction occur at all? And why should it be the same for all substances, regardless of what they're made of? There was no satisfying answer to these questions, and the Lorentz–Fitzgerald contraction seemed a very ad hoc way out of the Michelson–Morley contradiction.

Ingenious though the nineteenth-century physicists were, they weren't quite ingenious enough to break free of their nineteenth-century mindset and resolve the contradiction in a simple, fresh, and radical way. That resolution had to wait until 5 years into the next century, and when it came it was truly revolutionary yet profoundly simple.

EINSTEIN TO THE RESCUE

• • •

Albert Einstein was 8 years old when Michelson and Morley performed their 1887 experiment. Slow to speak, socially withdrawn, and stormy of temper, the young Einstein seemed not especially promising. But a magnetic compass his father showed him had evoked in 5-year-old Albert the first glimmers of his fascination with the deepest nature of physical reality. The compass needle's mysterious response to something unseen and unfelt made a lasting impression and stirred Einstein's lifelong search for the hidden principles governing the physical Universe.

Einstein: Approaching the Magic Year

Einstein the student got mixed reviews. In elementary school he was strong in logical subjects like mathematics and Latin, but showed little motivation for others that failed to captivate him. As an older student, Einstein's interests broadened to include philosophy, literature, and music. Always, though, he rebelled against rote learning, and eventually dropped out of his militaristically disciplined German high school to join his family who were then living in Italy. Without completing high school, and 2 years younger than the minimum age for admission, he took the entrance exam for the Zurich Polytechnic Institute. He did impressively well in mathematics and physics, but failed other subjects. Nevertheless, he was admitted with a year's delay, contingent on his gaining a high-school diploma. So Einstein enrolled in a high school in Aarau, Switzerland.

During his year at Aarau, at age 16, Einstein had a prescient insight that shows his mind was already stretching toward relativity. He wondered what would happen if one ran alongside a light wave, at its same speed. Obviously, one should see a stationary structure of electric and magnetic fields—just as a sailor moving with the speed of water waves would see, relative to the boat, stationary crests and troughs of water. The trouble was, Einstein realized, that a stationary electromagnetic wave structure is not consistent with Maxwell's electromagnetic theory. Much later Einstein remarked that 10 years' speculation on this quandary helped lead him to relativity.

In 1896, Einstein enrolled at the Zurich Polytechnic Institute. Here he continued his iconoclastic approach to education. Disgusted with a professor whose lectures were so outdated that they didn't include Maxwell's electromagnetism, Einstein skipped class and studied Maxwellian theory on his own. He was arrogant and confrontational to some of the physics faculty, who in turn found him insolent and lazy; they even suggested that he switch to medicine or literature. Einstein also found time for love with his fellow physics student Mileva Marić. She had come to Zurich to study medicine, then switched to physics—an unlikely choice for a woman in her day, but made possible by the liberal climate at Zurich, which had earlier graduated the first female PhDs in Europe. Marić's academic record was comparable to Einstein's, and in 1900 they and three classmates all took the final exam. Their grades were the lowest of the five; Einstein passed, but Marić didn't. Nevertheless, Einstein's letters reveal, they looked forward to a life of collaborative work in physics.

Success did not come easily. After graduation from Zurich, Einstein failed to find a permanent position. His classmates had all been offered faculty appointments at Zurich, but Einstein had so antagonized his professors that they would have none of him. Then Mileva discovered she was pregnant. Their daughter Lieserl was born early in 1902, but soon all trace of her vanished from the historical record. Did she die young? Did she live, perhaps never knowing herself to be Einstein's daughter? There's simply no evidence. Einstein, meanwhile, tried unsuccessfully to support his family as a private tutor. Eventually he secured a technical job in the Swiss Patent Office. This lowly position was actually good for Einstein; in addition to a stable income and contact with interesting technological ideas, it gave him plenty of time to pursue his own work in physics.

Einstein and Mileva Marić married in 1903. In 1904 their son Hans Albert was born. That brings us to 1905.

The Magic Year

Asked to picture Einstein, you probably conjure up images of an old man with wild hair and a frumpy sweater. But that's not the Einstein of relativity. In 1905 Einstein was a young father of 26 years, devoted to his family, to his work at the patent office, and to his physics (Figure 7.1). Despite his busy young life, lack of an academic position, and a paucity of colleagues, Einstein in 1905 nevertheless produced a burst of scientific advancement unprecedented except, perhaps, for Newton's plague years at Woolsthorpe.

Fig. 7.1 In 1905 Einstein was a young father, shown here with his infant son Hans Albert. (Courtesy of The Albert Einstein Archives, The Hebrew University of Jerusalem, Israel.)

In 1905 Einstein completed a total of six scientific papers. Three of these stand out as seminal works in the history of physics. The first paper introduced the photon, or quantum of light energy, and helped establish the fledgling science of quantum physics. That work earned Einstein the 1921 Nobel Prize in Physics. Two papers dealt with the size of molecules and provided the final convincing evidence that matter really does consist of atoms and molecules. The first of the molecular papers was Einstein's PhD dissertation for the Zurich Polytechnic; ironically, he submitted it only after the Zurich faculty had rejected his paper on relativity! The fifth paper of the year—a mere footnote to the theory of relativity—introduced the idea behind the famous equation $E = mc^2$.

Then there's the fourth paper. And a remarkable one it was. It cited no other scientific papers, contained a minimum of mathematics, and made few references to specific experiments. Yet with incisive clarity, ingeniously penetrating insight, and utter simplicity, Einstein in this paper resolved completely the contradictions posed by Michelson–Morley and other attempts to answer the question of Earth's motion.

Was this paper "The Theory of Relativity"? Yes, in content. No, in title. Rather, its modest title is "On the Electrodynamics of Moving Bodies." That title emphasizes the intimate connection between electromagnetism and relativity. Relativity grew out of contradictions arising from Maxwell's revelation that light consists of electromagnetic waves propagating at speed c. The natural question, Speed c relative to what? followed directly. The nineteenth-century answer, ether, then led to questions about Earth's motion. Recall that it also led to a perplexing dichotomy, in which one branch of physics—Newton's mechanics—obeys the relativity principle, meaning that the laws of mechanics are valid in all uniformly moving reference frames. But the other branch—Maxwell's electromagnetism—seemed to be valid only in the ether's reference frame and therefore did not obey the relativity principle.

Einstein's resolution was at once conservative, radical, and profoundly simple. It's stated in one brief sentence, called the Principle of Relativity:

> The laws of physics are the same in all
> uniformly moving reference frames.

That's it. This one sentence implies all of Einstein's special theory of relativity. It's conservative because it asserts for electromagnetism—indeed, for all of physics—what had been known for centuries about mechanics, namely, that there's no favored state of motion, no preferred frame of reference. If you look back in Chapter 3, where I introduced the principle of Galilean relativity, you'll find that Einstein's statement is identical except that the phrase "laws of motion" becomes generalized to "laws of physics." So in this sense Einstein's relativity is nothing new. It's just a generalization of Galilean relativity to all of physics—including electromagnetism. Einstein's relativity is clearly simple, stated in its entirety in just one brief sentence, but it's also radical. For Einstein to state it, and for you to understand it fully, requires dramatic restructuring of deep-seated commonsense notions about the nature of space and time. That is what the rest of this book is about, and it's what makes relativity seem a challenging, even mind-boggling subject. But I want to stress again and again that, in essence, relativity is simplicity itself. It's all based on the fact that there's no preferred state of motion for describing physical reality; that you and I, as long as we're both moving uniformly, will discover exactly the same underlying physical laws—even though we may be moving relative to each other. That's exactly the idea that I introduced in Chapter 2, when I convinced you that a tennis match and a microwave oven should work the same on Earth, on Venus, and on that planet in a distant galaxy moving away from Earth at nearly the speed of light. Motion—as long as it's uniform motion—simply doesn't matter. There's no one who can claim, "You're moving and I'm not." All states of motion—at least uniform motion—are equally valid. There's no such thing as being absolutely at rest or in motion. Only statements about relative motion make sense.

By the way, you might be questioning that caveat about uniform motion. Why shouldn't *all* states of motion be equivalent? We'll get there eventually, as Einstein did by 1915. But first things first: Einstein's 1905 relativity is the *special* theory of relativity. Special here doesn't mean that the theory is particularly great and wonderful (although it certainly is both!), but special in the sense of being specific and limited. Special relativity is limited to the special case of reference frames in uniform motion. Einstein's general relativity, which I'll introduce in Chapter 14, removes that limitation.

(Historically, Einstein actually presented relativity in the form of two postulates. The first was his assertion that the Principle of Relativity, which I stated above, applies to all of physics. The second, which Einstein noted "is only apparently irreconcilable" with the Principle of Relativity, asserts that the speed of light is the same in all uniformly moving reference frames. A more modern approach takes the view that the second postulate follows from the first; indeed, by 1910 physicists had shown rigorously that the second postulate is superfluous.)

Here's how the Principle of Relativity takes care of those pesky nineteenth-century dilemmas and contradictions. The Principle asserts that all the laws of physics are the same in all uniformly moving reference frames. Among the laws of physics are Maxwell's equations of electromagnetism. Those equations lead to the prediction that there should exist electromagnetic waves and that those waves should propagate at a particular speed—the speed of light, c. Speed c relative to what? Nineteenth-century physicists foundered around in the ether trying to answer this question, without success. But the Principle of Relativity provides a simple answer. Because the prediction of electromagnetic waves propagating at speed c is a prediction of the laws of physics, and because those laws are valid in all uniformly moving reference frames, it must be the case that electromagnetic waves propagate at speed c as measured in any uniformly moving reference frame.

Farewell to the Ether

Einstein's answer to the question, Speed c relative to what? renders the concept of ether unnecessary and, indeed, dangerously misleading. The ether was supposed to provide a medium in which light waves could propagate, and as such it would constitute a frame of reference that could claim to be truly at rest and in which alone the laws of electromagnetism would be valid. But the idea of a preferred frame runs afoul of the Principle of Relativity. If the laws of physics are valid in all uniformly moving reference frames, then the idea of a preferred ether frame has to go. So there is no ether. Relativity does away with it once and for all. As Einstein himself put it in his 1905 paper, "the introduction of . . . ether will prove to be super-fluous." Light propagates not through some material medium but through empty space. In that sense light waves are different from all other waves physicists had encountered, all of which require a physical medium—air for sound, water for water waves, rock for earthquake waves, and so forth.

Have Faith in Relativity!

There's a disturbing implication of relativity's assertion regarding the speed of light. Observers in different reference frames must still get the same value for the speed of light, *even though they're moving relative to each other*. That means you and I, measuring the speed of the very same light, get the same answer even though I'm at rest on Earth and you're whizzing by in a car, an airplane, or a high-speed rocket. It's that consequence—the invariance of the speed of light, even for observers who are moving relative to one another—that's so troubling and that's going to lead to a radical revision of your commonsense notions of space and time. We'll explore these ideas thoroughly in the next chapter.

For now, though, concentrate on where the invariance of c comes from. It results from nothing more than the Principle of Relativity, as applied to the laws of electromagnetism with their prediction of electromagnetic waves (including light) going at speed c. That's all. There's nothing hidden. Accept the Principle of Relativity—as you did intuitively with the tennis and microwave examples of Chapter 2—and the invariance of the speed of light follows. When you find yourself in disbelief at some consequence of relativity in subsequent chapters, just come back to this simple point. If you accept the Principle of Relativity, then it's just two steps to the invariance of c:

(1) The laws of electromagnetism predict electromagnetic waves going at speed c and (2) The laws of physics are valid in all reference frames; thus, the conclusion that electromagnetic waves go at c must be valid in all reference frames. From the invariance of c follow the many seemingly counterintuitive results of special relativity.

But why should you accept the Principle of Relativity? As a reader with a twenty-first-century perspective on the vastness of our expanding Universe, you readily bought my argument in Chapter 2 that the tennis match and the microwave oven should work the same on Earth, on Venus, and on that distant planet moving rapidly away from Earth. If you want to question that, ask yourself if you really believe Earth to be so special that only here, or, more accurately, only in a reference frame at rest with respect to Earth, are the laws of physics valid. In all the vast Universe, with countless galaxies each containing billions of stars and planets whizzing apart in the cosmic expansion, could our state of motion be so special?

Even if you want to ignore your post-Copernican instinct that Earth and its state of motion aren't special, look at the experimental evidence and the struggle of nineteenth-century science to fathom the question of Earth's motion. As I detailed in Chapters 5 and 6, the nineteenth century closed with no satisfactory answer to this question that is consistent with observations like aberration of starlight and experiments like Michelson–Morley. But Einstein's 1905 answer is consistent and simple: The laws of physics are the same in all uniformly moving reference frames. Period. Although conclusions drawn from Einstein's simple statement are going to upset your notions of space and time, I'll emphasize now and again later that those conclusions, while strange, are never contradictory. They follow from only one simple fact: The laws of physics are the same in any uniformly moving reference frame.

•

STRETCHING TIME

• • •

Why is relativity so hard to swallow? Because the invariance of the speed of light—a straightforward consequence of the Principle of Relativity—leads directly to what seem impossible conclusions. In this chapter I'll use the invariance of c in reaching these strange conclusions, but don't come away thinking that relativity is just about the behavior of light. Relativity is ultimately about the absence of any favored reference frame in the physical Universe, and all the strange conclusions we'll draw follow from that idea. The invariance of the speed of light is only one of many consequences of the relativity principle. The behavior of light provides a convenient stepping stone to relativity's deeper implications, but keep in mind that those implications reflect the fundamental nature of space and time, and aren't just about light.

Measuring the Speed of Light

Figure 8.1 shows you standing by a road. A traffic signal some distance away emits a flash of light, and the figure shows the light waves as they come by you. How fast? At speed c, of course. You could, in principle, verify that speed if I gave you a meter stick and a very accurate stopwatch. You would start the watch when the light flash first passed the front end of your meter stick and stop it when it passed the back end. Divide the distance—1 meter—by the

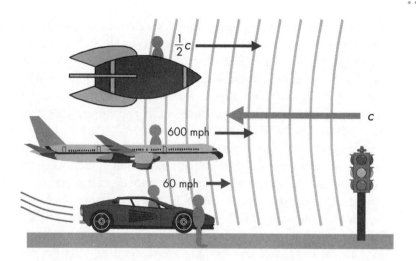

Fig. 8.1 Four observers each measure the same value c for the speed of light relative to themselves, even though they're in motion relative to one another.

measured time, and you'll have the speed of light. With a perfect meter stick, a perfect stopwatch, and perfect experimental technique, you'll get precisely the value c = 299 792 458 meters per second (this is the exact value for c; my "186,000 miles per second" and "300 million meters per second" are approximate). Of course you can't do this experiment with ordinary meter sticks and stopwatches, but modern electronics provides timing devices sufficiently fast and light pulses sufficiently short that one can easily measure c precisely over distances of a few meters—the size of a typical room.

Figure 8.1 also shows a friend driving by in a car at 60 miles per hour, toward the traffic signal. I've given your friend a perfect meter stick and a perfect stopwatch, identical to yours. Your friend uses exactly the same experimental technique on exactly the same light flash, to determine the speed of the flash. And what does she find? Because she's heading toward the traffic signal, you might guess that she'll measure a slightly higher speed than your 299 792 458 meters per second—higher by her speed of 60 miles per hour, or about 27 meters per second. But that answer isn't consistent with the Principle of Relativity, which states that the laws of physics are the same in all uniformly moving reference frames. One consequence of those laws is the existence of electromagnetic waves—including light— that propagate at speed c. Both you and your friend are in uniform motion, so that particular consequence of the laws of physics must hold for both of you. Both of you must measure the same value for the speed of light c. With perfect meter sticks and stopwatches, that value will be 299 792 458 meters per second.

Let me make it perfectly clear that I'm saying that each of you gets *exactly* the same value for c. You might think your speed relative to your friend is so small that you just don't notice the difference. So to make things more dramatic, Figure 8.1 shows two more observers, one in a jet plane going at 600 miles per hour and another in a rocket moving toward the traffic signal at half the speed of light. Surely an astronaut on the rocket sees the light approaching at $1.5c$. But no! That result would be inconsistent with relativity for the same reason it would be for the driver of the car. You, your friend, the airplane pilot, and the astronaut are all in uniform motion and, therefore, the laws of physics are equally valid for all of you. None of you—includ- ing yourself standing on the road—can claim in any absolute sense to be at rest. None of you can say that the laws of physics are cor- rect only for you. You're all in equally good situations—reference frames—for exploring physical reality, and one consequence is that you'll all measure precisely the same value for the speed of light, c.

But how can this be? How can different observers, moving rela- tive to one another, still measure the *same speed* for the *same light*? Ultimately the answer is simple: The laws of physics are the same for all observers in uniform motion, and the invariance of c follows directly from that principle. Whenever you find yourself asking, How can this be? about relativity, the answer always lies in the Prin- ciple of Relativity. Remind yourself how intuitive that principle seemed in Chapter 2, and how the experimental evidence and scien- tific quandaries of the nineteenth century led Einstein to affirm the relativity principle as the basis of his theory. If you still don't like what relativity has to say, then *you* answer the question, Speed c rel- ative to what? in a way that's consistent with experiments like Michelson–Morley and that doesn't put Earth, alone among all the cosmos, in a favored position.

A [False] Analogy with Sound

You might be tempted to use an analogy with sound to argue against my conclusion that all four observers in Figure 8.1 measure the same speed for light, so let's explore that analogy. Suppose we replace the traffic signal with a loudspeaker, a musical instrument, or a person shouting. Out comes a burst of sound waves, heading for the four observers. They measure the speed of this sound, each with the same meter stick and stopwatch technique (this is easier because sound goes much more slowly than light). Doesn't each observer get a different value for the speed of sound? Surely your friend in the car, heading toward the source of sound, measures a higher value than you do—higher by her speed of 60 miles per hour. Because a commercial jet's 600 miles per hour is nearly the speed of sound, the airplane pilot should measure almost twice the speed you do. And to the astronaut—woosh!—the sound should seem to be going by at about half the speed of light. So is that what happens? Do the different observers measure different speeds for the sound waves? If they do, doesn't that invalidate the conclusion that they'll all measure the same speed for light waves?

Yes, but no. The four observers all do measure different values for the speed of sound. Those values are, as you would have guessed, very nearly the sum of the official speed of sound and the speed of each vehicle. Why is that? Because sound moves at its "official" speed—about 700 miles per hour, or 340 meters per second—with respect to air, the medium in which sound is a wave. Think back to Chapter 4, when I introduced the idea of waves as disturbances of some medium. For sound waves that medium is air, so to say "the speed of sound is 340 meters per second" has an unambiguous meaning: it denotes the speed of sound waves relative to the air. If you're moving through the air, then you'll get a different value for the speed of sound *relative to you*—just as I suggested back in Figure 6.1. (By the way, the three "moving" observers need to stick their measuring instruments outside their vehicles to measure different speeds for sound because the air—the medium for sound waves—within each vehicle shares that vehicle's motion.)

But light doesn't work that way because it requires no medium. Light propagates through empty space, not through some material substance against which you can gauge your own state of motion. Nineteenth-century physicists, hung up on the concept of the ether, thought they had such a substance. If they had been correct, then ether would have been for light what air is for sound, and our four observers in Figure 8.1 would have measured different values for the speed of light. The only one to measure the "right" value would be that observer—if any—who happened to be at rest with respect to the ether. But the nineteenth-century physicists were wrong. They tried to measure differences in the speed of light associated with Earth's motion through the ether (that's what the Michelson–Morley experiment was all about), but they failed to find any difference whatsoever. Their failure led to seeming contradictions—contradictions that vanished with Einstein's abolition of the ether and his assertion that the relativity principle holds for all of physics. So light and sound do not behave analogously, and you can't argue against Figure 8.1 by analogy with sound.

Being Relativistically Correct

So what's going on? How do the observers in Figure 8.1 all get the same value for the speed of light? Although this follows from the relativity principle, to say so isn't a very satisfying explanation. There is an explanation but it requires stretching your commonsense notions of time and space.

One way out of our dilemma would be if something were wrong with some of the observers' measuring instruments. Maybe high-speed motion does something strange to meter sticks and/or stopwatches, something that just compensates for the motion and results in everyone measuring the same value for c. Take that spaceship going by at half the speed of light, a speed with which we have no first-hand experience. Might such rapid motion alter the behavior of clocks and meter sticks in a way that we don't notice in our slower, everyday motions?

This is a tempting explanation, and there is, in fact, a grain of truth in it. But in another sense it's dead wrong, in a way that obviously and dramatically violates the Principle of Relativity. Right here and now I want to make sure we agree to avoid any language that even hints at violation of that principle. So what's wrong with saying that high-speed motion affects measuring instruments? What's wrong is the implication that some instruments are moving at high speed and some aren't—and that the ones that aren't moving are more "right." That's bad because it violates the essence of the relativity principle, with its insistence that all observers in uniform motion have equal claim to be "right" when they experiment with the physical world. So it's wrong—relativistically incorrect—to say something like "The spaceship's clock reads differently because it's moving." One is often tempted to make such statements, and even some books on relativity routinely do so. But such a statement is always misleading because it implies that there's a frame of reference—not the spaceship's—that isn't moving. And it's precisely the existence of such a frame, with its claim to be absolutely at rest, that the relativity principle vociferously denies.

When talking about relativity, pledge to use only relativistically correct language! Watch out for phrases like "the moving clock," "the observer at rest," or "the high-speed spaceship." Unless it's clear what these things are moving or at rest with respect to, then the phrases are meaningless. Worse, they imply an absoluteness to motion and rest that is antithetical to the relativity principle. Regarding Figure 8.1, we might say, "The stopwatch in the space-ship reads differently from the one on Earth, because Earth and the spaceship are in relative motion," or, perfectly equivalently, "The stopwatch on Earth reads differently from the one on the spaceship because Earth is moving relative to the ship." Those phrasings are consistent with the relativity principle because they don't mention motion in an absolute sense. What isn't consistent is a statement like "The stopwatch in the spaceship reads differently from the one on Earth because the spaceship is moving, period."

By the way, you should be suspicious of words like "really" or "what actually happens" when you're talking relativity. Usually these have a hidden implication; "really" might mean "from the viewpoint of someone on Earth" and "what actually happens" might mean "as judged by someone who shares my frame of reference." That's relativistically incorrect because of the implicit assumption that some one point of view (frame of reference) has a unique claim to judge reality or what actually happens.

Suppose you still insist you're OK in claiming that the spaceship in Figure 8.1 is moving, period. Then you haven't yet accepted the truth of the relativity principle. If you do accept that principle, then you have to give observers on the ship exactly the same privilege: the spaceship's astronauts can claim that you, standing by the road, are the one who's moving. So who's right? Both, or neither. Better to avoid the issue altogether and talk only about relative motion. You: "The spaceship is moving relative to me." Astronaut: "You, and Earth, are moving relative to me." Both statements are correct, and neither implies any claim that anyone is at rest, or moving, in an absolute sense. Only relative motion matters. That's why it's called relativity.

I've summarized this point about relative motion in Figure 8.2, which shows two different stripped-down versions of Figure 8.1. Gone are the car and the airplane, to avoid clutter. Gone, too, is the Earth—to help overcome your natural prejudice that there's something special about the solid ground on which you, in Figure 8.1, are standing. Finally, gone is the traffic signal, because—as double-star observations show and as application of the relativity principle to electromagnetism makes clear—the speed of light does not depend on the relative motion of the light source and the observer measuring that speed. So there's nothing special about your situation just because you happen to be the one observer who's at rest relative to the traffic signal.

Figure 8.2a is drawn from your point of view, meaning that it shows the situation in your frame of reference. Relative to this frame, you're at rest and the spaceship is moving to the right at half the speed of light. You obviously measure c for the speed of the light. Your common sense might suggest that an astronaut on the spaceship should measure $1.5c$, but—hard as it may be to accept—the Principle of Relativity tells you otherwise. Figure 8.2b shows the situation in the ship's frame. Relative to this frame, the ship is at rest, and you're moving to the left at half the speed of light. (Really? Remember, only relative motion matters! Relative to an observer sitting at rest in the spaceship, you move rapidly away to the left.) An astronaut on the ship measures c for the speed of light. The astronaut's common sense

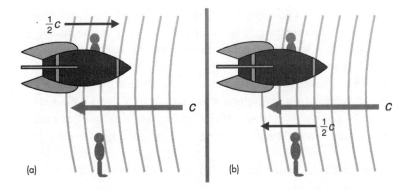

Fig. 8.2 Take away the ground and the traffic signal, and it's obvious that the roadside and rocket observers are in perfectly equivalent situations. Both think of themselves as being at rest, and both measure c for the speed of light. (a) From the roadside observer's viewpoint, the rocket is moving to the right at half the speed of light. (b) From the rocket's viewpoint, the roadside observer is moving to the left at half of c.

might suggest that you should measure only $0.5c$, since you're moving *away* from the light's source, but relativity tells the astronaut otherwise. Both you and the astronaut have equal claim to be in the right situation to do physical experiments, and if you believe relativity then each of you has to grant the other that same rightness.

Space and Time are Relative

Let's go back to Figure 8.1. How can all four observers measure the same speed c for light? I mentioned that there's a hint of correctness in the thought that maybe their measuring instruments read different things. Notice I didn't say here that anyone's instruments are wrong. This would imply that someone else's instruments are right, giving them the privileged position that the relativity principle denies. Nor am I saying that one set of instruments behaves differently from another; all work just the same way, timing the flash of light over a fixed distance. And all give exactly the same result: the ratio of the distance to the measured time is c, or 299 792 458 meters per second.

But how is this possible, given that the different sets of instruments are moving relative to one another? It's not that the instruments go "wrong," but something much deeper. It's the nature of time and space. Time and space are not absolute, but relative to one's frame of reference. Measures of time intervals and of spatial distances are simply different in different reference frames. And, of course, no one has claim to the right measures; all reference frames in uniform motion are perfectly acceptable for making valid measurements of space and time. It's the differences in space and time from one reference frame to another that ensure that all four observers in Figure 8.1 measure precisely the same value for the speed of light. The spaceship may be moving relative to you, but because its measures of space and time are different from yours, you both come up with the same value c.

Sense and Common Sense

How can time be different for different frames of reference? That sounds preposterous because your common sense suggests an absolute, universal time ticking off equal-length seconds everywhere and in every state of motion. It's as if there could be a master clock somewhere that establishes instantly a unique time throughout the whole Universe. Isaac Newton shared your commonsense concept of universal time: "Absolute, true, and mathematical time, of itself and from its own nature," he said, "flows equably without relation to anything external."* But where is that absolute, universal clock? In whose reference frame? Wouldn't such a clock establish a preferred state? And how would that sit with the relativity principle?

Your everyday experience may suggest that time is universal and absolute, but is your experience broad enough to establish that absoluteness beyond a doubt? No matter how sophisticated and

---

*Isaac Newton, *Principia* (Berkeley, CA: University of California Press, 1934); originally published in London in 1687 and trans. by Andrew Motte in 1729. This phrase appears near the beginning of the work, in the "Scholium," a discourse on time, space, and absolute versus relative motion that follows Newton's definitions of basic physical quantities.

well traveled you are, you've probably never moved faster than the 600 miles per hour of a commercial jet, *relative to anything important to you.* Even if you're a supersonic pilot or an astronaut, you've been limited to speeds of at most about 5 miles per second *relative to Earth.* That's still minuscule compared with the roughly 186,000-miles-per-second speed of light. (The italicized phrases are for relativistic correctness and also because you do move much faster relative to things that you don't directly perceive, such as that distant galaxy we discussed in Chapter 2 or the electrons zooming through your TV at 30 percent of *c* to create the picture on your TV screen.) So you lack the experience to know for certain that high relative speeds don't result in different measures of space and time. And at the low relative speeds you have experienced, maybe there are differences that are just too small for you to notice.

Your commonsense concepts of time and space were established very early in life. As a crawling baby, you explored space in its three dimensions. Nothing led you to believe that crawling faster relative to your surroundings in any way altered that space. Later, perhaps, came your notion of time. But by the age when you were left at day care, started school, or whined "Are we there yet?" from the back of the family car, you surely knew something about the passage of time. Again, nothing in your experience from then until now has suggested that the time between two events isn't an absolute quantity, the same for everyone.

But what if you had grown up crawling at 80 percent of the speed of light, relative to your immediate surroundings? Then you'd have no need of this book. Your commonsense understanding of time and space would be entirely consistent with the Principle of Relativity, and what presented a stiff intellectual challenge to no less than Albert Einstein would be intuitively obvious to you. It's just your provincialism—stuck on the hunk of rock called Earth, limited to speeds far less than *c* relative to your planet—that makes relativity hard to swallow. But relativity and your provincial common sense are not actually in contradiction. Your limited experience may lead you to think that time and space are absolute, the same for all who care to measure them. But your experience doesn't require that conclusion. It's equally consistent with space and time being relative to your frame of reference, given that all the frames of reference you've ever occupied have moved so slowly relative to one another that you don't notice the difference. Slowly, that is, in comparison with the speed of light.

Time Dilation

Exactly how do measures of time differ in different reference frames? Here I'm going to argue that difference qualitatively and present without proof how it works quantitatively. (Those who like math will find the details in the Appendix.) First, though, I need to clarify an idea that's already familiar to you—the idea of an *event*. Events play a major role in relativity, because they involve both time and space. An event is an occurrence, something that happens at a specific place and time. Your birth is an event; it occurred some*where* and at some *time*. Your reading this paragraph is another event; it, too, is happening some*where* and at some *time*. The two events are not the same, because they're separated in time and probably in space as well. (Although they could occur at the same place, at least in Earth's reference frame, if you happen to be reading in the very spot where you were born.) An event is completely specified by giving a *time* and a *place*. For example, the most recent major earthquake to strike the San Francisco Bay area resulted from a specific event—a slippage along the San Andreas fault. The event's location (place) was at Loma Prieta, in the Santa Cruz Mountains, 60 miles south of San Francisco and at a depth of 11 miles below Earth's surface. Its time was 5:04 PM PST on October 17, 1989. Another event, the landing of the NEAR spacecraft, marked humankind's first contact with an asteroid. This event occurred at time 12:02 PM PST on Monday, February 12, 2001, and its place was the asteroid Eros, then 196 million miles from Earth. Since every event has a time and a place, it makes sense to talk about the time intervals and spatial separations between any pair of events. For our two events, the time interval is approximately 11.28 years and the spatial separation is 196 million miles—as measured in a frame of reference at rest with respect to Earth.

I'm now going to convince you that the time interval between two events cannot be the same for two observers in motion relative to each other. The argument I'll use is based solidly in the Principle of Relativity and its consequence, the invariance of the speed of light. I'll use a rather artificial-seeming situation involving a bouncing light beam, but don't come away thinking this is only about light or about the particular situation I'll describe. Rather, it's about the nature of time. The example I present serves to illuminate what is, in fact, a universal aspect of time—namely, that measures of the time between two events differ in reference frames that are in relative motion.

So here's the situation, shown first in Figure 8.3a. There's a rectangular box with a light source at one end and a mirror at the other. A brief flash of light leaves the source. We'll call this occurrence—the departure of the light flash from the source—event A. The light travels up the box, hits the mirror, reflects, and returns to the source. We'll call event B the return of the light to the source. That is, event

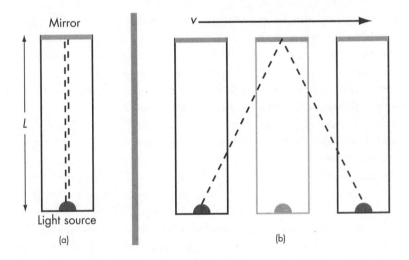

Fig. 8.3 A "light box," used as a clock to measure the time interval between two events. Event A is a light flash leaving the source. Event B is the light returning to the source, after reflecting off a mirror at the top of the box. Dashed line is the light path. (a) The situation in a reference frame at rest with respect to the box. The light travels a total distance twice the box length L. (b) The situation in a reference frame in which the box moves to the right at speed v. The light travels farther, but because it has the same speed c, the time between the two events must be longer.

B is completely specified by stating that it occurs at the bottom of the box (B's *place*) at the exact instant that the light flash reaches that place (B's *time*). The light's round-trip takes some time, which is the time between events A and B *as measured in a frame of reference at rest with respect to the box*. We need that clarification because the time between two events is not absolute but depends on one's reference frame. That's just what I'm trying to demonstrate.

You could calculate the time between the two events quite easily if you knew the length of the box. Let's call that length L. Then the light goes a total distance of twice L, because it makes a round-trip. And how fast does the light go? At speed c, as always. Knowing the distance and speed, it's simple to calculate the time—and I do that explicitly in the Appendix. But for this qualitative argument, all you need to know is that the light makes a round-trip of twice the box length and that it does so at speed c.

I warned you earlier to be suspicious of pictures in relativity books. Every picture is drawn from a particular point of view, that is, from the perspective of a particular reference frame. To understand what's going on, you need to know exactly what frame that is. In the case of Figure 8.3a, the reference frame of the picture is at rest with respect to the box, so it's in that reference frame that we've been discussing the time between events A and B.

Now let's look at the situation from another reference frame, one moving relative to the box. Equivalently, the box is moving relative to this new reference frame. So suppose the box is moving to the right, at some speed v, relative to the new frame (v here is for "velocity"). Figure 8.3b shows the situation in this reference frame. Because the box is moving relative to the reference frame, I've needed to show it in different locations at several different times. In particular, it's shown at the time of event A, when the light flash leaves the source; at the time of event B, when the light returns; and at an intermediate time when the light reflects off the mirror at the end of the box.

We want to know the time between events A and B as measured in the reference frame of Figure 8.3b. To find that time we need the distance the light travels between its departure from and its return to the source, and we need to know how fast it goes. Without doing any math, you can see that the light's path in Figure 8.3b is *longer* than it is in Figure 8.3a, which was drawn from the perspective of a reference frame where the box was at rest. That's because the light, relative to the reference frame of Figure 8.3b, takes a diagonal path heading to the mirror and another diagonal path coming back. Those diagonals are each necessarily longer than the length L of the box, since they incorporate both that length and some motion of the light sideways, along the direction of the box's relative motion. Again, there's nothing fishy about this. If I'm in a bus and I throw a ball straight up, it goes straight up and down relative to me. But if you're standing by the road you see the ball take a curved path that's longer than the path it takes in a reference frame at rest with respect to the bus. Similarly, the light takes a longer path in a reference frame relative to which the box is moving.

The only other thing we need to get the time between the light's departure and return—that is, the time between events A and B—is the light's speed. Here's where the relativity principle comes in. Again, one consequence of that principle is the invariance of the speed of light. So the light goes at c relative to the reference frame of Figure 8.3b. But it goes *farther* in this reference frame than it did in the frame of Figure 8.3a, at rest with respect to the box. Since it has the same speed, c, in both frames, it must *take a longer time* in the reference frame with the longer path—that is, in the reference frame relative to which the box is moving. So we're forced to conclude that the time between events A and B in a reference frame in which the box is at rest is *shorter* than the time between the same two events in a reference frame with respect to which the box is moving.

Let me be clear that I'm not saying simply that the light took two different trips, one of them longer, and therefore that the longer trip took a longer time. That would be obvious without relativity. Rather, the *same* light took *one and the same trip*, and we examined that trip from the viewpoints of two different reference frames. The beginning and end of that trip are the events A and B, and observers in both frames agree about what those events are and that they indeed mark the endpoints of the light's trip. What they disagree about is the time between those *same* events.

That's all well and good, you might say, in this highly artificial situation where the two events in question happen to involve a light beam that must go at c in any reference frame. Surely, you say, this has nothing to do with everyday events of the sort you and I experience. But it has everything to do with such events. In principle, I could have set up a "light box" at any event you care to name—for example, at your birth—and arranged for the flash to depart at the instant of that event. By moving the light box at the right speed I could arrange for it to arrive at a later event of your choosing—for example, your reading this chapter. And by choosing the length of the box (it would have to be very long for your birth and reading events!), I could arrange for the light to return to the source just as the box arrived at where you are as you read this chapter. All the arguments I made from Figure 8.3 would apply here, and we're forced to conclude that the time between the events of your birth and your reading this chapter is different in different reference frames. This is about *time*, not about light beams and mirrors. The light box serves as a device for exploring the nature of time and for concluding that the time between two events is relative to one's frame of reference. But the light box doesn't cause that difference; the difference is intrinsic in the nature of time. Measurement of the time interval between two events with ordinary clocks would, in principle, reveal exactly the same discrepancies between different reference frames, regardless of the presence or absence of the light-box device. In fact, all good clocks in a given reference frame will measure the same time between two events, and that time will be different from that measured by identical clocks in another reference frame in relative motion. By a "good clock," I mean any device that accurately measures the passage of time. That includes the light box in Figure 8.3, an atomic clock, your bedside alarm clock, or your wristwatch. It also means time-dependent processes like the vibration of a radio wave, the beating of your heart, or even the biological clocks that govern the aging of your body. All measure the same underlying phenomenon—time itself.

Before exploring further this strange new understanding of time, let's tie up a few loose ends. First, be very clear where and how the Principle of Relativity entered my argument. It entered in the assumption that the speed of light was c in *both* reference frames. Again, that assumption follows directly from the relativity principle as applied to electromagnetism. Had we not made that assumption, you could have argued that the light in Figure 8.3b was going faster than in Figure 8.3a because it shared the box's horizontal motion. Then, even though it was going farther, it would have taken the same time in both frames. But you can't make that argument for the same reason that you can't argue for different speeds of light for the observers in Figure 8.1. It's the Principle of Relativity that rules here, not some commonsense but incorrect notions you've developed about how light ought to behave. And the relativity principle implies that the speed of light is the same in all uniformly moving reference frames.

There's a more subtle point you might want to challenge. In arguing that the light's path in Figure 8.3b is longer than in Figure 8.3a, I've implicitly assumed that the box length L is the same in both reference frames. But I've mentioned repeatedly that relativity affects both *time* and *space*. So how do I know the box length isn't different in a way that shortens the path in Figure 8.3b and thus voids the time difference? It is true that measures of space, too, depend on reference frame (we'll soon see just how) but that only happens for measures taken along the direction of relative motion between two frames. There's no difference for measures, like that of the box length L in Figure 8.3, that are perpendicular to the direction of relative motion. You don't have to take this on faith, because it follows directly from the relativity principle. However, I'd like to hold that argument until we've explored further our new discovery about the relativity of time.

"Moving Clocks Run Slow"

The shortening of the time interval between our events A and B in the light-box frame is known as *time dilation*. It's as if clocks—and indeed, all manifestations of time—are running more slowly in the light-box reference frame than they are in the frame relative to which the box is moving. Many books on relativity summarize time dilation with the phrase "moving clocks run slow." This is a dreadful phrase, as relativistically incorrect as they come! (It's also grammatically incorrect; the last word should be the adverb "slowly.") So what's wrong with "moving clocks run slow"? Here's what: Who's to say that one clock is moving and another isn't? Given the relativity principle, the phrase "moving clocks" is utterly meaningless and worse, since it applies an absoluteness to motion that is antithetical to the very essence of relativity.

Here's a better, albeit wordier, description of time dilation:

> The time between two events is shortest when measured in a reference frame where the two events occur at the same place.

This description is fully consistent with the light-box example that led us to time dilation. In the light-box reference frame, events A and B do occur at the same place, namely the bottom of the box. The same place? Even though the box is moving? If you're thinking that way, then you're not yet fully accepting relativity. The box's reference frame is a perfectly good one for doing physics, and no one can make an absolute claim that the box is moving. To someone in the box, it's at rest. An observer sitting in the bottom of the box when the light flash goes off (event A) can stay in the same place and will later be present when the flash returns (event B). Since the observer doesn't have to move relative to the box, the two events occur, for this observer, at the same place. If you still want to claim that the two events aren't "really" at the same place because the box is "really" moving, then you're granting a special status to the other reference frame, the one relative to which the box is moving. If you insist on doing so, then you're in violation of the Principle of Relativity. Watch out for that word "really"!

You might also object by citing a more concrete example of two events, namely your birth and your reading this chapter. Suppose you were born in San Francisco and you're now sitting in New York. Am I saying that San Francisco and New York are in the same place? Of course not. But to an observer sitting in a car that left San Francisco at your birth and just now arriving in New York after a slow but steady trip, the *events* of your birth and of your reading occur at the same place, namely right at the car. Note the distinction here between *events* and *places*. San Francisco and New York are indeed in different places, and different observers agree about that (although, as you'll soon see, they disagree about how far apart they are). But as measured by an observer in a reference frame moving with respect to Earth, the positions of the two cities are continually changing. By choosing just the right motion relative to Earth, our carbound observer can arrange for those positions to be the same— namely, right at the car—at both your birth and your reading of this chapter. That's what I mean when I say that two events occur at the same place in some reference frame.

Figure 8.4 illustrates the meaning of my description of the time-dilation phenomenon. The picture is drawn in the reference frame of two clocks, C_1 and C_2, that are located at different places. A third

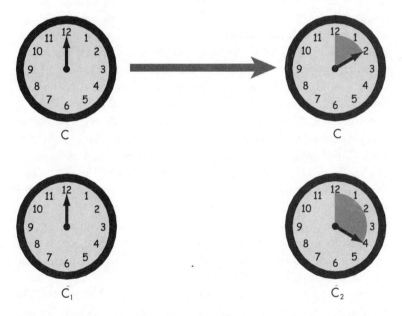

Fig. 8.4 Two clocks, C_1 and C_2, are at rest with respect to each other, and the figure is drawn in the reference frame of these two clocks. A third clock, C, moves relative to C_1 and C_2 in a direction that takes it first past C_1, then on past C_2. Clocks C and C_1 read the same time when the two coincide; this is event A. When C reaches C_2 (event B), C reads less elapsed time than C_2. Thus, the time between events A and B is shorter in C's frame of reference. (Here C's speed is such that its time is half that measured by C_1 and C_2.)

clock, C, moves relative to the others in such a way that it passes first C_1 and then C_2. We'll assume that C passes very close to the other clocks. Then just as C passes C_1 we have a distinct event, occurring at a definite place and a definite time. We'll call that event A. Event B will be C passing C_2. In the reference frame of clocks C_1 and C_2, the two events obviously occur at different places, namely, the distinct locations of the two clocks. But clock C is moving relative to the others in such a way that it's present at both events. That is, events A and B occur at the same place in C's reference frame—in just the same way that your birth and reading occurred at the same place in the reference frame of the car in the previous paragraph. Then, says time dilation, the time interval between events A and B will be shorter as measured on clock C than as measured by clocks C_1 and C_2. I've indicated this time difference on the clocks in Figure 8.4.

There's an important sense in which the situations in the two reference frames involved in our time-dilation examples are not equivalent. In one frame—the frame of the light box in Figure 8.3 or of clock C in Figure 8.4—it takes only one clock to measure the time interval between the events in question. But in the other frame it necessarily requires two clocks. That's because the events occur at the same place in the first frame but at different places in the second. (You might ask why you can't use a single clock in place of clocks C_1 and C_2, and watch for the occurrence of the two events from afar. You could do that, but then you'd have to compensate for the travel time of light coming from the two events to your observing location. You already know enough relativity to realize that such compensation is going to require some care. In any event, an observer right at clock C doesn't need to apply any such compensation, so the two situations still wouldn't be equivalent.)

There's another issue in using two separated clocks to measure the time interval between two events. Obviously, those two clocks have to be synchronized—meaning they both read the same time at the same instant. That's not hard to achieve, but it does require a little thought. When clock C_1 reads exactly noon, for example, it could send out a light flash. When the light flash reaches clock C_2, an observer there could set C_2 to read noon *plus the time it took the light to travel from C_1 to C_2*. Then the clocks would be synchronized. Now, an observer looking at the two clocks wouldn't *see* them reading exactly the same time unless the observer were equal distances from both clocks. An observer closer to one clock would *see* that one reading a later time, since light from that clock would have left more recently because of the shorter distance. But knowing the distances to the two clocks that observer could nevertheless infer that they were, in fact, synchronized. The observer would thus *observe* that the clocks were synchronized even though she doesn't *see* them reading the same time.

Note here the important distinction between *seeing* and *observing*. Seeing, when what you see involves objects at different distances, doesn't give an accurate picture of what *is*, because of the different travel times for light from the two objects. (When I say "is" here, I mean what "is" for an observer in your reference frame; I'm not referring to some absolute truth that's independent of reference frame. Thus clocks C_1 and C_2 are synchronized *to an observer at rest with respect to the two clocks*, even if that observer doesn't *see* them reading the same time.) When I say that someone observes a given situation, I'll always mean that that person determines what actually happens as judged from his or her frame of reference. Often such observation means more than just looking; instead, it means looking and then compensating for light's travel time.

You might think I'm making a big fuss over not much, because light travels so fast that the differences in travel time from different objects are negligible. That's true for most everyday situations. But when objects move relative to you at very high speeds, or when they're very far away, the differences become significant. Those are precisely the situations in which relativistic effects are evident and in which your commonsense notions of time and space become inadequate. As a concrete example, consider astronomers trying to take a snapshot of galaxies in a particular region of the sky. The nearest galaxies might be so close that their light takes only a few million years to reach us. For the most distant, that time becomes billions of years. So the astronomers' image is not a snapshot at all, in that it doesn't show the galaxies as they *are*, or even as they all *were* at some common time in the past, but rather as they *were* at times that vary dramatically from galaxy to galaxy. What the astronomers *see* is not what *is*.

Getting Quantitative

Just how significant is the time difference between events as measured in different reference frames? That depends on the speed of the frames' relative motion. For relative speeds small compared with the speed of light, time dilation is not at all obvious and is very difficult to measure. For relative speeds approaching c, though, the effect becomes dramatic.

Although I'm taking a nonmathematical approach to relativity in this book, I nevertheless want to show you quantitatively the effect of time dilation. One simple formula sums up that effect. If you like math, you can follow its derivation in the Appendix. Even if you aren't into math, you'll recall that sometime in high school or earlier you learned how to use the Pythagorean theorem to find the diagonal of a right triangle. That involved taking the sum of the squares of the two shorter sides, then taking the square root. What's this got to do with time dilation? Simply this: the light path in Figure 8.3b forms the diagonals of two right triangles. As the Appendix shows rigorously, finding the light travel time in the reference frame of Figure 8.3b thus involves the Pythagorean theorem. So it's no surprise that the formula for time dilation involves squares and square roots:

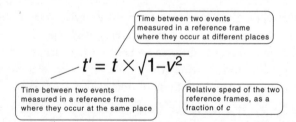

I've labeled this time-dilation formula to make it clear exactly what each term means. The formula gives the time, t' (read "t prime"), between two events A and B, as measured in the reference frame in which two events occur at the same place (e.g., the frame of clock C in Figure 8.4). To the right of the equal sign, t is the time as measured in a reference frame in which the two events occur at different places (e.g., the frame of clocks C_1 and C_2 in Figure 8.4). The symbol v under the square-root stands for the relative speed of the two reference frames, given as a fraction of the speed of light. That is, $v = 0.5$ means half the speed of light, $v = 0.9$ means 90 percent of the speed of light, and so forth. So what the formula says is that the time t' in the frame where the events occur at the same place is given by the time in the other frame, multiplied by the square root of $1 - v^2$. Let's take a closer look at this quantity. If $v = 0$, then the two frames aren't in relative motion and we're left with the square root of 1, or just 1. So our formula gives $t' = t$. Of course: In this case the two frames are really one and the same reference frame, and the formula is just telling us that observers in the same reference frame all agree on the time between events. But if the two frames are in relative motion, then the two times are no longer equal. Consider first a fairly low relative speed, say 10 percent of the speed of light, so $v = 0.1$. Then $v^2 = 0.01$, and the quantity $\sqrt{1 - v^2}$ becomes the square root of 0.99, which is very nearly 1. So even in this case—a relative speed of nearly 19,000 miles per second—the two times are still very close. It's only when the relative speed becomes a significant fraction of c that time dilation becomes substantial. With $v = 0.8$ for the relative speed between two frames, for example, you can convince yourself that $\sqrt{1 - v^2} = 0.6$, meaning that the time between two events measured in a frame where the events occur at the same place is only a little over half what it is in the other frame. At $v = 0.99$, the square root is about 0.14, and the one time is only about a seventh of the other. Finally, you might wonder about the case $v = 1$, corresponding to relative motion at the speed of light. Here the formula gives $t' = 0$. This suggests that time does not pass at all in the frame where the two events occur at the same place! But as we'll see in Chapter 12, a relative speed of c is not possible for frames of reference associated with physical objects like human observers or clocks of any kind.

Wrapping It Up

The crucial point of this chapter is that measures of the time between events need not be the same in two reference frames in relative motion. In particular, observers in two different reference frames measure different time intervals between the same two events—with the shortest time measured by an observer for whom the two events occur at the same place. Although I spent a long time elaborating on this point, it follows directly from the Principle of Relativity and its consequence, the invariance of the speed of light. If you accept the Principle of Relativity, then you can't logically escape this conclusion about the relativity of time.

But is this strange new behavior of time, as embodied quantitatively in the time-dilation formula, at all relevant to anything? Can we find or imagine situations in which it occurs and is important? Or is all this just an academic exercise?

As the numerical examples in the preceding section show, we can expect obvious time-dilation effects only when relative speeds are very high—close to the speed of light. We might detect time dilation at lower relative speeds, but then only with very sensitive experiments. At everyday relative speeds, including those of jet aircraft and even today's spaceflight, time dilation is simply too small for us to notice directly.

Even so, time dilation does occur and is measurable. The effect shows up dramatically in experiments involving subatomic particles moving, relative to Earth, at speeds approaching c. As I outlined in Chapter 1, it's even been measured in clocks flown around Earth in ordinary aircraft—although here the effect is minuscule. We can imagine a future with high-speed space travel, where observers on Earth and passengers in spacecraft would measure very different times between the same events. We'll explore these realities and possibilities in the next chapter.

STAR TRIPS AND
SQUEEZED SPACE

•   •   •

Suppose we have a spaceship capable of traveling, relative to Earth, at 80 percent of the speed of light. We'll dismiss the obvious technical challenges in building such a craft, and we'll also assume that it can jump essentially instantaneously from being at rest on Earth to its cruising speed of $0.8c$. Such an acceleration would surely be fatal, but we'll imagine we've somehow overcome this problem. Finally, we'll equip the ship with the latest instrumentation, including an accurate clock just like those used on Earth.

Years and Light-Years

Before we set out on a star trip, I want to introduce a way of describing interstellar distances that makes a lot more sense than feet and miles, or meters and kilometers. Instead of those familiar units, we'll measure distances in light-years. Despite the word "year" in its name, the light-year is a unit of *distance*, not time. One light-year is, simply, the distance light travels in a year. How far is that? It doesn't really matter, but you can find out approximately, if you want, by multiplying 186,000 miles per second by the number of seconds in a year (about 30 million). The result is about 6 trillion miles. It's more meaningful to compare the light-year with real distances. For example, the nearest stars beyond our Solar System are a few light-years away. Our Milky Way galaxy measures about

100,000 light-years across, and we're about 30,000 light-years from its center. The Milky Way's nearest large neighbor, the great spiral galaxy in the constellation Andromeda, and the most distant object visible with the naked eye, is about 2 million light-years away. The most remote objects astronomers have spotted with the Hubble Space Telescope are quasars some 12 billion light-years distant. So the light-year is a convenient unit when we're talking interstellar or intergalactic distances. But it's too big to be useful within our own Solar System; the Sun, for example, is only about 8 light-minutes from Earth, and the Moon is just over 1 light-second away. Either of these distances is only a tiny fraction of a light-year.

In describing travel in our hypothetical spaceship, then, we'll use the light-year as our unit of distance and we'll use the year as our measure of time. Finally, it would be nice to express the all-important speed of light in these new units. That's easy! Because a light-year is defined as the distance light travels in 1 year, the speed of light is, by definition, 1 light-year per year. So our $0.8c$ spacecraft goes 0.8 light-years per year relative to Earth. Its speed, as measured in light-years per year, is also precisely the number v—the speed as a fraction of the speed of light—that we're supposed to use in our time-dilation formula. So working in light-years and years is going to make things easier.

Star Trip! One-Way

Suppose you set out in that $0.8c$ spaceship to visit a star 20 light-years from Earth—as measured in Earth's frame of reference. We'll assume Earth and star are essentially at rest relative to each other (a reasonable assumption, given that nearby stars don't move very fast relative to Earth). Then Earth and star are in the same reference frame, which I'll call the Earth–star frame. We'll suppose that there are identical clocks located on Earth, at the star, and in the spaceship, and that the Earth and star clocks are synchronized as described in the preceding chapter. So the Earth and star clocks are like the clocks C_1 and C_2 in Figure 8.4, and the ship clock is like clock C in that figure.

How long does the star trip take? For the Earth–star frame the answer is easy: We know the Earth–star distance (20 light-years) and we know the speed. Back in elementary school you learned that *distance = speed × time*; after all, that's just what speed means, namely, how far you go in a given time. Therefore *time = distance ÷ speed*. So the trip time as measured in the Earth–star frame is

$$t = (20 \text{ light-years}) \div (0.8 \text{ light-years/year}) = 25 \text{ years.}$$

Does this make sense? Sure: If the ship were going at the speed of light (1 light-year per year) it would take exactly 20 years to make the 20 light-year journey. It's going a little slower, so the trip should take a little longer. If all the clocks read 0 years when the ship starts out, then our answer for *t* shows that the Earth and star clocks will read 25 years when the ship reaches the star.

What about the ship clock? Like clock C in Figure 8.4, this is a clock for which the two events of interest—the ship's departure from Earth and its arrival at the star—occur at the same place. So it's going to take less time as measured on the ship clock than it does on the Earth and star clocks—a conclusion forced on us by the argument of the previous chapter, an argument based on nothing more than the Principle of Relativity. The time-dilation formula we developed in the previous chapter applies to just such a clock as our ship clock, so we can work out the time *t'* as measured on the ship:

$$t' = t \times \sqrt{1 - v^2} = 25 \text{ years} \times \sqrt{1 - 0.8^2} = 25 \text{ years} \times \sqrt{1 - 0.64}$$
$$= 25 \text{ years} \times \sqrt{0.36} = 25 \text{ years} \times 0.6 = 15 \text{ years.}$$

(You could do the math on a calculator, but I showed all the details because with $v = 0.8$ the square root works out so nicely to give 0.6 for the time-dilation factor.) Figure 9.1 summarizes your star trip and the different times involved.

The difference here is substantial! In the Earth–star frame, the time between the ship's departure and arrival is 25 years. In the ship's frame, it's only 15 years—for a difference of 10 full years. Like everything that's come before, this conclusion follows inescapably from the Principle of Relativity. But what does it mean,

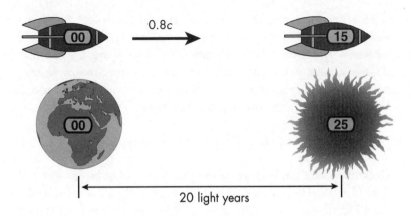

Fig. 9.1 At 0.8*c*, the trip from Earth to star takes 25 years in the Earth–star reference frame but only 15 years in the spaceship's frame. Times are shown on three digital clocks, one on the ship (shown twice) and one each on Earth and star. Figure, including clock readings, is a composite of two different times: first when the ship passes Earth and then when it passes the star.

besides that the ship's clock seems to "run slow" compared with the Earth–star clocks? In particular, what happens to you as you travel on the ship? Does the trip seem to take just 15 years, or is it "really" a 25-year trip in a spaceship whose clock somehow isn't reading right? If you're thinking the latter, then you're being relativistically incorrect! The ship's frame is just as good as the Earth–star frame for exploring physical reality. What happens in the ship's frame is every bit as real and valid as what happens on Earth. It isn't just that the ship's clock *reads* 15 years—it's that, in the ship's frame of reference, only 15 years of time have elapsed between departure from Earth and arrival at the star. *All* manifestations of time reflect this interval. In particular, you arrive at the star 15 years older than you were when you left Earth—not the 25 years older that your earthbound friends might expect.

Would you be aware, while traveling, that you're aging more slowly than usual? Would you feel your heartbeat slow down and notice your hair growing more slowly? Absolutely not! To you, in your spaceship, everything seems perfectly normal. It must! Why? Because you're in a perfectly good frame of reference for exploring physical reality. The laws of physics work just as well for you as they do on Earth. If you felt something strange because of the ship's motion, something you didn't feel on Earth, then you could genuinely say that there was something unusual about the spaceship's reference frame, something that wasn't true of the Earth's frame—namely, that the ship was moving in some absolute sense while Earth wasn't. But, of course, it's just such specialness of one reference frame over another that relativity denies. So everything—including the passage of time in all its manifestations from atoms to clocks to human bodies—seems perfectly normal onboard the spaceship.

At this point you're probably full of questions. What if you radio back to Earth periodically, maybe at each birthday? Will your friends back home recognize that you're aging more slowly than they? Suppose some friends your own age had traveled earlier to the star, via a much slower spaceship. When you stop at the star and join them, will you really be younger? Just what is it that's holding back time on the spaceship? What's getting into every clock, into the aging mechanism of each cell in your body, and into every electron whirling about every atom, to slow them all down?

I'll answer all these questions eventually, but for now I want to address just the last one. The answer is that nothing is holding things back. Don't go looking for a mechanism that slows everything down. To do so is to say that there's really a universal time but that something unusual happens in the moving spaceship to make all time-consuming processes take longer. That's no good because, again, it singles out Earth's frame as the one in which this time-slowing mechanism isn't at work, and it identifies the ship's frame as one in which something unusual happens because the ship is moving. To say those things about Earth and spaceship is to violate the relativity principle.

There's one subtle point I need to make. We've applied the idea of time dilation—that the time between the same two events is different when measured in reference frames in relative motion—to the duration of your star trip. We can do that if both reference frames, namely the Earth–star frame and the ship frame, are indeed in uniform motion. That's because the Principle of Relativity says that the laws of physics are equally valid in all reference frames *in uniform motion*. If the ship starts from rest relative to Earth, then travels to the star and slams on its brakes, it's certainly not moving uniformly while it starts and stops. The reason I've assumed the ship can accelerate instantaneously to 0.8*c* relative to Earth, and then stop abruptly at the star, is so that it spends essentially the entire journey in uniform motion. To be completely correct, we should identify the departure event as occurring just after the ship has accelerated. Since the acceleration takes place instantaneously, the ship is still right at Earth but it's now moving relative to Earth. Its clock still reads 0 because the acceleration took essentially no time. The ship then moves, uniformly, relative to Earth and star, until the arrival event, when it's at the star but hasn't yet applied its brakes. Between the departure and arrival events so defined, the ship indeed moves uniformly, and we can apply the Principle of Relativity and its logical consequence, the time-dilation formula, to reach our conclusion that the time between departure and arrival is 10 years less for those riding the ship than for those remaining on Earth or at the star.

A cleaner way to present the star trip would have been to imagine an alien being in a spacecraft that happens to come whizzing by Earth at a steady 0.8*c*. Later the alien passes a star that, as measured in the Earth–star frame, is 20 light-years from Earth. We can draw exactly the same conclusion in this case as we did before, namely, that the alien measures 15 years between passing Earth and passing the star, while observers with synchronized clocks at Earth and star report a 25-year interval between the time those on Earth see the craft go by and when observers at the star see it pass. What's simpler about this situation is that we don't have to think about the alien spacecraft starting and stopping; it's truly in uniform motion relative to Earth the whole time, so it's completely obvious that the relativity principle applies. But I emphasize again that it also applies to your star trip, at least during the interval when you're moving uniformly relative to Earth. Later in this chapter we'll reconsider the effects of starting and stopping.

Squeezing Space

How can the spaceship get from Earth to star in only 15 years? After all, the distance is 20 light-years, so even light should take 20 years.

As you're probably aware, and as I'll elaborate on in Chapter 12, no material object can go faster than light. So how can the spaceship get to the star in only 15 years?

The answer lies in the fact that measures of space, as well as of time, are different in different reference frames. Let's look at things from your viewpoint as you ride the spaceship. You know you're moving at $0.8c$ relative to Earth and star, and you know that your journey takes 15 years. So how far have you gone? You're in a perfectly good frame of reference for doing physics, so the elementary-school formula *distance = speed × time* works just fine for you. So to you, the distance you've traveled from Earth to star must be

$$d' = (0.8 \text{ light-years/year}) \times (15 \text{ years}) = 12 \text{ light-years}.$$

(I've called this distance d' ["d prime"] for consistency with my earlier notation, where I used t' ["t prime"] to represent time measured in the ship frame.)

There's no inconsistency here and no faster-than-light travel. To you in the spaceship, the Earth–star distance is 12 light-years, and at 0.8 light-years per year, it makes perfect sense that your trip takes a little over 12 years—15 years, to be precise. What's troubling is that something you thought was objectively real and absolute, namely the distance between two objects, simply isn't. Like measures of time, measures of space also depend on one's frame of reference.

Note, by the way, that the 12-light-year distance in the ship frame is precisely 60 percent of the 20-light-year value that the Earth–star distance has in the Earth–star frame. That 60 percent, or 0.6, is just the relativistic factor $\sqrt{1 - v^2}$ that we calculated for time dilation with $v = 0.8c$. In fact, that's generally true. If the distance between two objects is d in a frame of reference where the two are at rest, then in a reference frame moving at speed v relative to the objects, the distance will be contracted by this same factor, giving

$$d' = d \times \sqrt{1 - v^2}.$$

For this *length contraction* to occur, the relative motion must be along a line between the two objects. If the motion is perpendicular to that line, then there's no contraction; if it's at an angle, then the contraction factor is somewhere between.

I referred to the shrinking of the distance between Earth and star, or any to other objects at rest with respect to each other, as length contraction. That's because the two objects could be the opposite ends of a single physical thing, like a ruler. To someone moving relative to the ruler, it's contracted to a shorter length. The ruler may still say "12 inches," but as measured by an observer moving relative to the ruler, it won't measure 12 inches. But isn't the ruler really 12 inches long? To ask that is to suppose there's one special frame in which measures of length or distance are correct and that therefore they're wrong in other frames. Obviously, that view violates the relativity principle. On the other hand, one's relationship with an object like a ruler is certainly simplest in a frame of reference at rest with respect to the object. For that reason the length measured when one is at rest with respect to an object is called the object's *proper length*. Here "proper" doesn't mean "correct" as much as it does "proprietary"—in the sense of "belonging" to the object, or intrinsic to the object in its own frame of reference. But that doesn't mean the object's proper length has any transcending reality. Observers in different reference frames will measure different lengths for an object, and they'll all be correct. Measures of space and time just aren't absolute. An object is longest in a reference frame where it's at rest and shorter in any other frame. Just how short depends on one's speed relative to the object, as given by the formula above.

Once again I need to remind you of the difference between *observing* and *seeing*. When you observe an object moving relative to you, you measure its length by, perhaps, noting where its front and back ends are in relation to a ruler or meter stick at rest in your reference frame. Since the object is moving, this can be a bit tricky. You have to be careful to note the positions of the front and back ends *at the same time*. But you can do that, and the result will be the contracted length of the object as measured in your reference frame. However, that doesn't mean you'll *see* the object contracted. That's because light from different parts of the object reaches you at different times, and those time differences are significant for an object moving fast enough relative to you that its length is noticeably contracted. Remarkably, the object *appears* not contracted but rotated! However, I'm not going into the details of how this comes about because it's a bit of a distraction from the bigger theme of the nature of space and time.

Finally, a historical note. The length-contraction formula $d' = d \times \sqrt{1 - v^2}$ is precisely the remedy Lorentz and Fitzgerald proposed in the late 1800s to resolve the quandary of the Michelson–Morley experiment. If the Michelson–Morley apparatus contracted by this amount in the direction of its motion through the ether, Lorentz and Fitzgerald correctly argued, then the travel times for light along the two legs of the apparatus would remain the same, and the experiment wouldn't be able to detect Earth's motion through the ether. So Lorentz and Fitzgerald got it partly right, in that they correctly predicted a motion-induced contraction of material objects. But they remained philosophically mired in a relativistically incorrect way of thinking, because for them the contraction occurred against a background of absolute space and time. Theirs was a contraction of material objects in an uncontracted space. The relativistically correct interpretation of length contraction is that measures of space itself differ in different reference frames and that differing measures for the length of material objects reflect this underlying relativity of space.

It Really Happens!

Does all this really happen? Does time really pass more slowly on your spaceship than it does on Earth and star? Is the Earth–star distance really less for you as you ride the spaceship? The answer to these questions is yes—although if you're in a deeply relativistic mode of thinking then these questions themselves may be bothersome. Why, after all, can't you on the spaceship consider Earth to be moving, and therefore its clock to "run slow"? You can, and I'll return to this point later. For now, though, let's consider how we might show that it really happens. Unfortunately, we don't yet have the technology to carry out the star trip, so we can't directly experience time dilation and length contraction. We do have access to objects that move relative to us at very high speeds but those objects are too small for us to ride on, because they're subatomic particles.

Surprisingly, though, they carry "clocks" that allow us to confirm quite dramatically the effects of time dilation.

Here I'm going to describe an experiment that was done in the 1960s with the express purpose of showing directly that time dilation and length contraction do occur.* Special relativity had been thoroughly verified long before the 1960s, so this experiment was designed and performed especially to be convincing to folks other than scientists. I'm going to introduce the real experiment first by imagining a fictitious but analogous experiment closely related to our Earth–star trip.

Imagine I have some unusual clocks with the remarkable property that they self-destruct in 20 years, exploding into a jumble of hands and gears. These aren't very accurate clocks, so some explode after only 18 actual years, a very few as early as 12 or 15 years, while a few last 25 or 30 years. If I take 100 new clocks and wait 20 years, roughly half of them will be gone. If I wait 25 years, only a very few will be left. But if I wait only 15 years, nearly all of them will still be around.

Let's load your spaceship with 100 of these clocks, and send you off on that star trip at 0.8c. Your friends remain on Earth, although a few of them have already traveled—again, on a much slower spaceship—and are waiting at the star. When the ship reaches the star, what will you or your star-based friends find? Will the spaceship be full of clocks or will most of them have self-destructed before the journey is over? Your Earth–star friends, who agree that your trip takes 25 years, might expect to find very few clocks. But for you the trip takes only 15 years, so nearly all the clocks should still be around. So which is it? This is something we can test objectively by opening up the ship when it reaches the star and looking for the clocks. If we don't find many, then we can conclude that the trip time was really 25 years, even for the traveling clocks. If we do find a lot of clocks remaining, then we're forced to conclude that the trip time as judged by these clocks was considerably less than the 20 years it takes before half the clocks will explode.

*The experiment is summarized by D. H. Frisch and J. H. Smith in *American Journal of Physics* 31 (1963): 242–355, and in their film *Time Dilation—An Experiment with mu-Mesons* (Educational Development Center, 1963).

If we did this experiment, which of course is no more possible than the original star trip with a single accurate clock onboard the ship, then relativity says that we should indeed find lots of clocks, because in the ship frame the trip time is only 15 years.

In fact, an almost identical experiment is possible, and it's the one that was done in the 1960s. Instead of clocks, the experiment uses subatomic particles that are radioactive—meaning that they self-destruct by exploding into several other particles. The particular particles are called *muons*, and they're formed when cosmic rays slam into atoms high in Earth's atmosphere. A steady rain of these muons comes downward through the atmosphere at speeds approaching *c*. The experiment consists of two parts. First, experimenters high on a mountaintop (Mount Washington, New Hampshire) use a muon detector to determine the intensity of the "muon rain" (Figure 9.2a). The detector is designed to catch only muons moving with a narrow range of speeds very close to $0.995c$. The experimenters detect just over 500 such muons each hour at the mountaintop. Now the time it takes muons to self-destruct is well known, although the 1960s experimenters actually measure it again

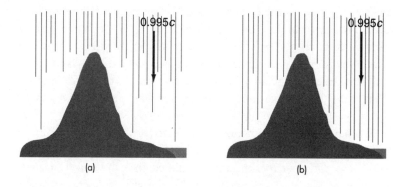

(a)　　　　　　　　　　　(b)

Fig. 9.2 Muons are radioactive particles that act like clocks. They arrive in great numbers at the top of Mount Washington. (a) They self-destruct at such a rate that very few would survive to reach sea level, if time in the muons' reference frame were the same as on Earth. (b) The muons experience time dilation, and thus most of them reach sea level. Detection of these many muons confirms time dilation.

as part of their experiment. That self-destruct time (analogous to the 20 years for our self-destructing clocks) is short enough that even muons moving at $0.995c$ would seem very unlikely to make it from the altitude of Mount Washington down to sea level without self-destructing on the way.

Now, that 500 muons per hour figure isn't unique to Mount Washington; we would measure very nearly the same quantity each hour anywhere within a few hundred miles if we were at the same 6,300-foot altitude as Mount Washington. So here's what the experimenters do: They go down to Cambridge, Massachusetts, at sea level, and set up the same muon detector, again looking for muons moving at $0.995c$. They know that overhead, some 6,300 feet up, there are 500 such muons every hour coming into a region the size of their detector. If they didn't believe in relativity, then the experimenters should expect to find very few muons at sea level, as shown in Figure 9.2a.

Now let's look at things from the muons' point of view. They're like the clocks in our spaceship, which are present both as the ship passes Earth and again when it reaches the star. The muons are present as they pass the 6,300-foot altitude of Mount Washington, and—if they haven't self-destructed—they're present again at sea level. So they're clocks that are present at two events, and for them the time between those events should be shortened according to our time-dilation formula. For a speed of $0.995c$, that formula gives a relativistic factor $\sqrt{1-v^2}$ equal to one-ninth. So the time to get from Mount Washington to sea level, as judged by the muons, should be only one-ninth what an observer at rest on Earth would expect. That muon time is so much shorter that very few muons should decay on the way from 6,300 feet to sea level.

We have a clear distinction between two possible outcomes. If relativity is correct, and time dilation occurs, then very few muons will decay and the experimenters should catch many muons in their detector at sea level (Figure 9.2b). If relativity isn't correct, then most of the muons should decay on the way down, and the sea-level experiment should find very few (Figure 9.2a). So what happens? The result is clear: Nearly as many muons reach sea level each hour as arrive at the top of Mount Washington. The muons are clearly experiencing time dilation, which at their speed relative to Earth is a dramatic effect. A look at the actual numbers in relation to the muons' lifetime not only confirms that time dilation occurs but also that it obeys precisely our time-dilation formula. Time for the muons elapses at one-ninth the rate it does as measured by two clocks at rest on Earth, one on the mountaintop and one at sea level.

We can also ask the same question here that we did for the spaceship. How can the muons possibly get from Mount Washington's altitude to sea level in so short a time that they don't self-destruct on the way? Doesn't that require them to be moving faster than light? No! Again, length contraction comes to the rescue. Just as measures of time are different in the muons' reference frame, so are measures of space. For the muons, the height of Mount Washington is contracted by the same factor (one-ninth), so the 6,300-foot mountain is, to the muons, only 700 feet high.

You may still be dissatisfied, because subatomic muons are so outside the range of your everyday experience. But the muon experiment really is completely analogous to our star trip, and we have every reason to believe that what holds for the muons would hold for ordinary clocks, people, and spaceships. If that weren't the case, then we would have the Principle of Relativity applying in the subatomic realm of physics but not to the physics of ordinary objects. Think back to Chapter 2, and you should find that possibility very unsettling.

Still, it would be nice if we could measure time dilation in ordinary clocks big enough for us to see and hold. Well, we can—but the effect is less dramatic because we can't get regular-sized clocks to speeds that, relative to us, are anywhere near the speed of light. Furthermore, because the effect of time dilation is so small at the relative speeds we can achieve, we need clocks that are good enough to mark extremely small time differences. We have such clocks—they're atomic clocks like those used to set the world time standards. These clocks are accurate to a few billionths of a second. In an experiment aimed at confirming relativistic effects on time, atomic clocks that initially read the same time ended up disagreeing by several hundred billionths of a second as a result of relative motion at the speeds of commercial aircraft. We'll revisit and elaborate on this experiment at the end of the next section.

A Round-Trip Star Trip—The Twins Paradox

Making our star trip a round-trip brings us to one of the most famous results of special relativity—the so-called twins paradox. Again we have the same trip to a star 20 light-years distant, as measured in the Earth–star frame. Now you board the spaceship and zoom off at $0.8c$. Your twin sister remains behind on Earth. When you reach the star you immediately turn around and zoom back to Earth, again at $0.8c$. You land, climb out of the ship, and greet your sister. How do your ages compare?

We've already done the numbers for the outgoing trip: According to clocks on Earth and star, the one-way trip to the star at $0.8c$ takes 25 years. According to the ship clock, time dilation gives a one-way trip of only 15 years. What about the return trip? It's identical to the outgoing trip. We still have the situation described in Figure 9.1, except that the ship is going in the opposite direction. Nothing in our careful analysis in Chapter 8 suggested that the direction mattered. There we found that a clock that's present at two events reads less elapsed time between those events than do two synchronized clocks in a frame of reference where the events occur at different places. That's as much the situation for the return trip as it was for the outgoing trip. For the return, the two events are departure from the star and arrival at Earth. They occur at the same place in the ship's frame (right at the ship, since you're present at both events even though you never move from your seat) and at different places in the Earth–star frame. So the time-dilation formula applies just as much to the return trip, which we conclude takes 15 years of ship time but 25 years of Earth–star time. Then the round-trip takes 30 years of ship time but 50 years of Earth–star time. So, when you return to greet your twin sister, you're 20 years younger than she is!

This result is unambiguous. You're standing right next to your sister and all can see that she's much older. There's no question of someone observing the two of you having to compensate for different light travel times or other complications, because you're right there beside each other. You really are 20 years younger than your sister, and no one can see it any other way.

By the way, things are a bit more ambiguous with the one-way trip, because there's the problem of communicating over the Earth–star distance to establish your relative ages. If we send advance observers to the star, there's still the complication of time dilation affecting them—which is why I had them sent well in advance in a much slower spaceship. We could, with care, establish an unambiguous age difference if you travel to the star, stop, and report your age. But it's much clearer if we consider the round-trip, because then you end up standing right next to someone who did nothing but stay at rest relative to Earth while you were traveling.

Why is this result paradoxical? It's still unsettling to think that you can end up 20 years younger than your twin, but if you really accept the relativity principle then you're going to have to accept that age difference as well. The paradox comes from considering the round-trip as viewed from the reference frame of the spaceship. To you, onboard the ship, the trip starts with Earth receding into the distance at $0.8c$. Later, just as you're at the star, Earth turns around and comes back toward you. When the trip is finished you're again standing next to your twin on Earth. From your point of view it looks like you stayed at rest in the ship, while Earth and your twin went off on a round-trip at $0.8c$. So why isn't she the one who's younger?

The resolution of this seeming paradox lies in what's *special* about the special theory of relativity. Remember that special relativity applies only to the special case of reference frames *in uniform motion.* Remember even further back to the philosophical shift from Aristotelian thinking to the ideas of Galileo and Newton. For Aristotle motion was really absolute and to keep something moving required a force. For Galileo and Newton, uniform motion became a natural state of affairs, requiring no force or other cause. For Galileo and Newton, it's *changes* in motion that are important. Newton's laws quantify this new philosophical view, showing just how forces cause *changes* in motion. Einstein's special relativity builds on this Newtonian idea, extending to all of physics the view that no one can make an absolute claim to be moving or at rest. This is the essence of the special relativity principle: that the laws of physics are the same no matter how you're moving—*as long as you're moving uniformly.*

Now to the extent that Earth is in uniform motion—and to a very good approximation, it is—your earthbound twin is in a reference frame in uniform motion and the laws of physics work just fine for her. As long as you're in the spaceship moving uniformly, they work just fine for you, too. But when the spaceship turns around at the star, and when it starts and stops at Earth, its motion is decidedly not uniform. You, the ship's passenger, know that because you feel the force of your seatback accelerating you as the ship leaves Earth, you experience the strong forces needed to turn you around at the star, and you feel the ship braking when it returns to Earth. Your twin back on Earth feels nothing as the ship accelerates, as it turns around, and as it stops. So there really is a difference, in that the ship's motion *changes* and Earth's doesn't. It's this difference that invalidates the argument from the ship's perspective. The ship is not in a single uniformly moving reference frame during the entire round-trip, so we can't expect the laws of physics to apply, in the spaceship frame, to the entire journey. We can apply them to the individual phases when the ship is in uniform motion, and that's how we conclude, correctly, that the ship time on each leg of the trip is only 15 years. But we can't say the ship is in uniform motion for the entire trip or that Earth makes a round-trip including a turnaround when the ship is at the star. The ship really does turn around, and Earth really doesn't!

Do you sense an absoluteness about motion creeping in here? Yes, in that I'm saying *changes* in motion are absolute in the context of special relativity. *Motion* itself isn't absolute but *changes* in motion are. Everyone agrees that you feel strong forces when your spaceship turns around and that your twin back on Earth doesn't. The presence of those forces isn't a relative thing; they and the resulting change in the ship's motion have an objective reality that transcends one's frame of reference. The difference between Earth and ship perspectives on the round-trip is real, and it means that you and your twin really are in different situations. Your argument that your twin sister should be younger is simply invalid, because you're applying the results of special relativity in the nonuniformly moving reference frame of your spaceship, where the physics of special relativity is not valid. So there's no paradox.

You might wonder why uniform motion is so special, and how we can know for sure that a reference frame really is in uniform motion. Einstein wondered too, and that helped lead him to his general theory of relativity—a theory that removes special relativity's restriction to uniform motion. We'll explore the general theory in the final three chapters and will revisit the twins paradox in the context of general relativity.

Time Travel!

Our round-trip at $0.8c$ to that star 20 light-years away resulted in a 20-year age difference between the traveling twin and the earthbound twin. Could we travel in a way that would result in even greater age difference?

Suppose we can crank our spaceship up to a speed arbitrarily close to c. Then the round-trip to the 20-light-year-distant star will take almost as short a time as it would take for light—namely 40 years, round-trip. That's the time as measured in the Earth–star frame. But as the ship's speed relative to Earth and star approaches c, the speed v in the relativistic factor $\sqrt{1-v^2}$ approaches 1 and the factor $\sqrt{1-v^2}$ itself approaches 0. So the ship time, $t' = t \times \sqrt{1-v^2}$, approaches 0 as well. If you're in the ship, you can make that time an hour, a minute, a second, or whatever you want, provided you set your speed close enough to c. So the greatest age difference between you and your stay-at-home twin would be just over 40 years. You get that maximum difference if you go so close to c that (1) the time in the Earth–star frame is essentially the same as the 40-year round-trip time for light itself and (2) the factor $\sqrt{1-v^2}$, and therefore the ship time, is very close to 0. In that case your star trip amounts, for you on the spaceship, to a nearly instantaneous 40-year jump into the future on Earth!

Forty years is the best you can do with a trip to that 20-light-year-distant star. What if you go farther? For a star 50 light-years distant, the shortest possible round-trip time, as measured by observers on Earth, is just over 100 years. Again, just how long it takes in ship time depends on exactly how close your speed is to c. You can make that time 1 year, 1 week, 1 day, 1 second, or whatever you want. If you make it very short, then a round-trip to the

50-light-year-distant star amounts to a jump 100 years into the future on Earth. So if you're really curious to know what things are going to be like on Earth a century from now, you can find out, without getting significantly older, by taking a round-trip at nearly the speed of light to a point 50 light-years distant. Time travel into the future really is possible!

But there's a catch. If you don't like the Earth you find 100 years in the future, there's no going back. You're either stuck where (or, rather, when) you are, or you can take your chances on a jump further into the future. But you can't go back. That would cause all sorts of con-tradictions, as you know if you've seen the film *Back to the Future*.

You can play this time-travel game with any numbers you like. The farther you go in space, the farther you can jump into the future. A round-trip to the center of our galaxy, 30,000 light-years distant, takes a minimum of just over 60,000 years in the Earth's frame of reference. But on a spaceship making that trip, the round-trip could take a few years, a day, an hour, or even less, depending on just how close v in $t' = t \times \sqrt{1-v^2}$ is to 1. So with a trip to the galactic center you can jump 60,000 years into the future. If you travel at nearly c to the Andromeda galaxy, 2 million light-years away, then you'll return to Earth 4 million years in the future.

Could it really happen? If you accept the Principle of Relativity, you have to say yes. Experiments with muons and atomic clocks confirm time dilation, the relativistic phenomenon at the basis of this futuristic time travel. Some day real people, using real space-craft, may travel seemingly impossible distances in what are never-theless minuscule times. If they choose to return to Earth, they'll have time traveled decades, centuries, millennia, or even further into the future!

THE SAME TIME?

• • •

At this point you may still be a bit uncomfortable with relativity and its strange consequences like different-aged twins and time travel to the future. But I hope you see the logic that takes us directly from the Principle of Relativity to these unusual implications. The relativity principle itself should be firmly grounded in your mind; after all, it made complete sense in Chapter 2, and by Chapter 7 it had emerged from the quandaries of late-nineteenth-century physics as the one clear way to resolve the muddle of motion and ether.

A Test of Faith

Now I'm going to put your faith in relativity to the test. Consider once again a one-way trip to that distant star. To make things really simple, forget about starting at Earth and stopping at the star. Rather, the spaceship just comes zooming past Earth at a steady 0.8c, then later zooms past the star. In other words, the ship is always in uniform motion, so it's always a good reference frame in which to do physics.

The preceding chapter should have convinced you that, from the viewpoint of observers on Earth, time on the spaceship "runs slow." Specifically, the time between the event of the ship passing Earth and the later event of the ship passing the star is shorter in the ship's frame of reference than it is in the Earth–star frame.

Now here's my question: What do you, a passenger on the space-ship, have to say about clocks in the Earth–star frame? To a friend on Earth, your clock runs slow. Do you want to say, then, that from your shipbound perspective, Earth and star clocks must run fast? Sure sounds like a logical answer. After all, if your clock is running slow compared with clocks on Earth and star, then surely Earth and star clocks must be running fast compared with yours.

But wait! I've set things up in this example so that you in the spaceship and your friend on Earth really are in equivalent situations, namely, the state of uniform motion. Your friend says, "I'm at rest on Earth, and the spaceship is moving relative to me, so the ship's clock is running slow compared with my clocks." But your situation is per-fectly equivalent, so you must be able to say the analogous thing: "I'm at rest in the spaceship, and Earth is moving relative to me, so Earth's clocks are running slow compared with mine." Each observer—you in the spaceship and your friend on Earth—must claim that the other's clocks are running slow. How can that possibly be? Yet if you really accept relativity—that the laws governing physical reality are the same in all uniformly moving reference frames—then you're forced to this seemingly absurd conclusion.

You might still object that the spaceship is "really moving" while Earth isn't, so the situations aren't really equivalent. But by now you're surely past such relativistically incorrect thinking! Unlike the round-trip we considered in the preceding chapter, or even a one-way trip with starts and stops, the scenario here really has both ship and Earth in completely equivalent situations. Neither changes its motion in any way, so both are in the state of unchanging, uniform motion that makes them both suitable reference frames for applying the laws of physics. Or maybe you want to say that the ship is moving in one direction relative to Earth, and Earth is moving in the opposite direc-tion relative to the ship, so that makes their situations different. But nothing in our arguments leading to time dilation depended on the direction of the relative motion. I could have redrawn Figure 8.3b in a reference frame in which the light box was moving to the left, and the argument from that figure would not have changed one bit. So the direction of relative motion can't matter.

So which is it? Do Earth's clocks run slow or fast from the viewpoint of the ship's clock? The answer, if we accept relativity, has to be that they run slow. To understand this answer, we have to find our way out of the contradiction it seems to imply—the contradiction that, relative to Earth, the ship clocks run slow while at the same time, relative to the ship, the Earth and star clocks run slow. Escaping this contradiction will give you further insights into the nature of time and especially the status of simultaneous events, that is, pairs of events that occur at the same time.

Simultaneity Is Relative!

What does it mean to say that two events occur at the same time? If the two events also occur at the same place, then there's no question. If we *see* the events simultaneously, then they must have occurred simultaneously. But suppose the events occur at different places. Then an observer watching the two events has to compensate for the time it takes light to get from each event to determine if they were, in fact, simultaneous. You've already seen enough of relativity to know that the constancy of the speed of light might make that compensation different for observers in relative motion. So if one observer determines that two events are simultaneous, it becomes an open question whether another observer, moving relative to the first, will deem them simultaneous.

Another approach to determining simultaneity is to set up clocks at the locations of the two events. At each clock, station a reporter who will report to you the time of the event occurring at the location of that clock. If both times are the same, then you can claim that the events were simultaneous. For this to work, you have to be sure those clocks are synchronized. What does that mean? It means both clocks read the same time, say noon, at exactly the same instant. In other words, the event of one clock's hands pointing to noon is simultaneous with the other clock's hands pointing to noon. You could synchronize the clocks by standing midway between them and sending a light flash to your reporters at the clocks. Each reporter sets the clock at the instant the light flash arrives. But again we're back to judging simultaneous events by methods that involve sending light signals through space and, again, relativity leaves open the question of whether observers moving relative to you will agree that the events are simultaneous. In this case, that means such observers might not agree with you that your clocks are synchronized.

In fact, events that are simultaneous in one frame of reference may not be simultaneous in another reference frame. I'm now going to demonstrate this rigorously in a way that follows from the Principle of Relativity. To do so, I'll invoke relativistic length contraction—a result which, as I showed in the preceding chapter, follows logically from the relativity principle. Recall that an object is longest in a reference frame in which it's at rest and shorter as measured in any other reference frame. In our earlier examples, one "object" in question was the Earth–star pair, whose separation contracted from 20 light-years in the Earth–star frame to 12 light-years in the ship frame. Another object we considered was Mount Washington, 6,300 feet high in the Earth frame but contracted to only 700 feet in the reference frame of muons for which Mount Washington was moving at $0.995c$.

I'm now going to consider two distinct objects and look at how length contraction applies to each of them in different reference frames. The objects are two identical airplanes, and they're flying toward each other at a substantial fraction of the speed of light (these are no ordinary airplanes!). The airplanes pass, one just above the other. I want to consider two events associated with this passing. The first event will be the nose of the upper airplane passing the tail of the lower one. We'll call this event A. The second event will be the tail of the upper airplane passing the nose of the lower one. This is event B. We want to know whether events A and B are simultaneous.

Figure 10.1 shows the situation as observed in a reference frame in which the two planes approach with the same speed. Obviously, neither plane is at rest in this reference frame, so each is shorter, because of length contraction, than it would be in a reference frame where it was at rest. Because both are moving at the same speed relative to the reference frame of Figure 10.1, each is contracted by the same amount. So in this reference frame the two planes have the same length. When they're alongside each other, as in Figure 10.1b, the nose of the upper plane coincides with the tail of the lower one *at the same time* that the tail of the upper plane coincides with the

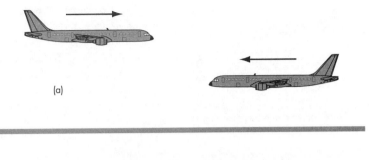

(a)

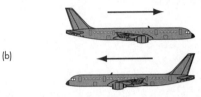

(b)

Fig. 10.1 (a) Two identical airplanes approach. The figure is shown in a reference frame in which the planes approach with the same speed. Event A is the nose of the upper plane passing the tail of the lower plane; event B is the tail of the upper plane passing the nose of the lower one. (b) In this reference frame, events A and B are clearly simultaneous.

nose of the lower one. That is, *events A and B are simultaneous in the reference frame of Figure 10.1.*

Now consider a reference frame in which the upper plane is at rest. In this frame the upper plane has its greatest length. But the lower plane is moving *faster* relative to the upper plane than it was relative to the reference frame of Figure 10.1, so in the rest frame of the upper plane, the lower plane is *even shorter* than it was in Figure 10.1. Figure 10.2a shows the situation at the time the nose of the upper plane coincides with the tail of the lower plane—that is, at the time of event A. Figure 10.2b shows the situation a little while later. The upper plane hasn't moved, since this is its rest frame. But now the nose of the lower plane has reached the tail of the upper plane. This is precisely our event B, so Figure 10.2b shows the situation at the time of event B. The conclusion is obvious: In the rest frame of the upper plane, the frame of Figure 10.2, *event A occurs before event B.*

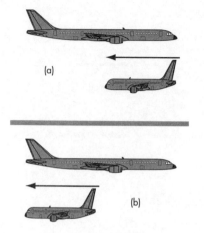

Fig. 10.2 In a reference frame in which the upper plane is at rest, that plane is longer than in Figure 10.1 because it isn't length-contracted. The lower plane, however, is moving faster relative to this frame than it was in the frame of Figure 10.1, so it's even shorter. (a) The situation at the instant of event A; (b) event B. Clearly event A precedes event B in this reference frame.

Finally, Figure 10.3 shows a reference frame in which the lower plane is at rest. Now the lower plane is longer, the upper one shorter, and it's obvious that in this reference frame *event B occurs before event A*.

The example of the airplanes shows clearly that *events that are simultaneous in one reference frame may not be simultaneous in another reference frame moving relative to the first*. Comparison of Figures 10.2 and 10.3 shows something even harder to swallow: the order of two events can be different in different reference frames.

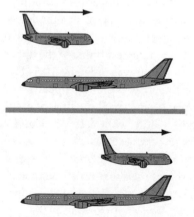

Fig. 10.3 Like Figure 10.2, but now in the reference frame in which the lower plane is at rest. Here event B comes first. Incidentally, Figures 10.1, 10.2, and 10.3 are scaled so that the speed of the two planes relative to the reference frame of Figure 10.1 is 0.6c.

That might seem particularly disturbing if you think about events that are causally related—like your birth and your now reading this book. Are there really observers for whom the reading comes before the birth? Fortunately, not. Causality is a big enough issue that I'll devote the entire next chapter to it; for now, let me just assure you that not all pairs of events can have their time order reversed. All we've learned from the airplane example is that those events that are simultaneous in some reference frames may not be simultaneous in other frames, and it's precisely such events whose time order itself depends on the observer's frame of reference.

By the way, I've been saying that events simultaneous in one frame "may not be simultaneous" in another frame. Why "may not" as opposed to "are not"? Are there cases where events can be simultaneous in different frames? Yes, if the relative motion of the two frames is perpendicular to the line joining the two events. For example, observers moving from the bottom toward the top of the page in Figure 10.1 will judge events A and B to be simultaneous, no matter how fast they're moving relative to the frame of Figure 10.1. For them, contraction of the airplanes squeezes them in the vertical direction, not the horizontal, so it affects both planes the same way.

"Moving Clocks Run Slow," Revisited

I now want to use your newfound knowledge about simultaneous events to resolve the apparent contradiction in my assertion that earthbound observers find that the spaceship clocks run slow, while observers on the ship find that Earth's clocks run slow. Look again at Figure 10.2, the airplane passings viewed from the reference frame of the upper plane. In this frame the lower plane's tail passes the upper plane's nose (event A) *before* the lower plane's nose passes the upper plane's tail (event B). Here's the point: in this particular situation, in the reference frame of the upper plane, it's the event that takes place further to the right—that is, further in the direction from which the "moving" plane is coming—that happens first.

Now, let's apply this understanding to our star trip, where we found that the trip time measured in the spaceship's frame was shorter than the trip time measured in the Earth–star frame. In the ship frame, we timed the trip with a single clock in the spaceship. In the Earth–star frame, we timed it with two different clocks, one located at Earth and the other at the star. Of course, those clocks had to be synchronized, meaning that they both read 0 years at the same instant. To be more precise, in the Earth–star frame the event of the Earth clock reading 0 years is simultaneous with the event of the star clock reading 0 years. Figure 10.4a shows all three clocks at the instant the ship passes Earth, at which instant all three read 0 years. You can tell that the figure is drawn from the viewpoint of the Earth–star frame, because it's in that frame that the events of the Earth clock reading 0 and the star clock reading 0 are simultaneous. But we've just found that events simultaneous in one reference frame aren't simultaneous in another, so those events can't be simultaneous in the ship frame. What, then, is their time relationship?

The ship frame is like the frame of the upper airplane in Figure 10.2. Ship and airplane are both at rest in their respective frames, and objects of interest (the other plane, or the Earth and star) are moving toward them from the right. Events that are simultaneous in one reference frame (the frame of Figure 10.1 in the airplane example or the Earth–star frame in our star-trip example) are not simultaneous in another frame (the frame of the upper plane or the ship frame) and, furthermore, the event that is farther to the right occurs first. That means the event of the star clock reading 0 occurs *before* the Earth clock reads 0, as observed from the ship's frame of reference. In other words, the star clock is *ahead* of the Earth clock. So when the Earth clock reads 0 time, the star clock reads a *later* time. That's exactly what Figure 10.5a shows. Here, as observed in the ship frame, Earth (heading to the left) passes the ship at the instant the ship and Earth clocks both read 0. But the star clock is ahead, so it reads a later time. I won't go through the math, but that later time is, in fact, 16 years.

You can probably begin to see now how we're going to get out of the contradiction that each observer thinks the others' clocks run slow. In the ship frame the one-way trip takes 15 years. But I'm claiming that observers on the ship must see Earth and star clocks running slow—and by the same relativistic factor $\sqrt{1-v^2} = 0.6$

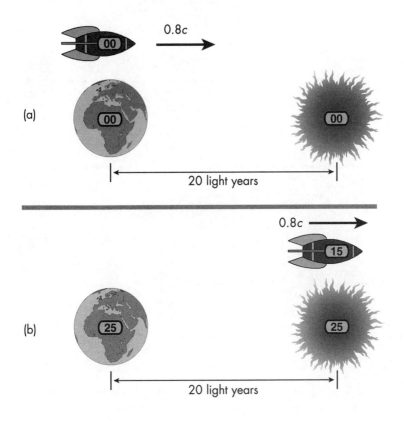

Fig. 10.4 Star trip of Figure 9.1, shown at two different times: (a) the situation at the instant the ship passes Earth; (b) its arrival at the star. Figure is drawn from the Earth–star frame of reference.

because the speed of Earth relative to the ship is the same $0.8c$ as that of the ship relative to Earth. So to the ship's observers the clocks on Earth and star advance only $(0.6) \times (15 \text{ years}) = 9$ years. When the ship reaches the star (or, from the ship's viewpoint, the star reaches the ship), the star clock has advanced only 9 years. *But,* as observed in the ship's frame, the star clock already read 16 years

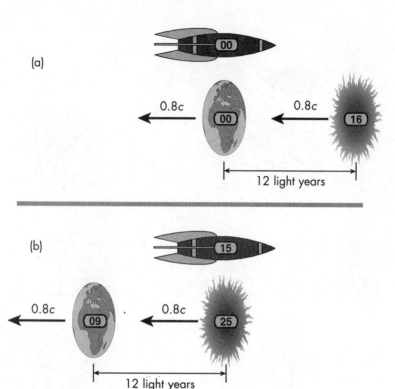

Fig. 10.5 Star trip, now drawn from the ship's reference frame. Here the ship is at rest, while the Earth and star move to the left at 0.8c. Note that the ship is longer in its own frame, while Earth, star, and the distance between them are contracted. (a) As the ship passes Earth, both ship clock and Earth clock read 0. In the ship frame the Earth and star clocks aren't synchronized, and the star clock is, in fact, 16 years ahead. (b) The ship clock advances 15 years between the times Earth and star pass the ship, just as in Figure 10.4, but Earth and star clocks are "running slow," and they advance only 9 years. Because it was 16 years ahead, the star clock still reads 25 years as the star passes the ship, just as in Figure 10.4.

when the ship passed Earth. So now it reads 16 + 9 = 25 years—precisely the trip time as measured in the Earth–star frame.

Let's sort all this out. Turn back to Figure 10.4, where (a) shows the situation in the Earth–star frame when the ship passes Earth and

(b) shows the situation in this frame when the ship passes the star. In this frame all the clocks read 0 as the ship passes Earth (Figure 10.4a). Later, as the ship passes the star, its clock reads 15 years while Earth and star clocks—which are synchronized in the Earth–star frame—both read 25 years (Figure 10.4b). The interpretation of the clock readings in the Earth–star frame is that 25 years have elapsed in that frame but only 15 years have elapsed in the ship frame because the ship clock runs slow.

Now look again at Figure 10.5, where (a) shows the situation in the ship frame when Earth passes the ship and (b) shows the situation in this frame when the star passes the ship. As in the Earth–star frame, both ship clock and Earth clock read 0 as Earth and ship pass. But in the ship frame, the clocks in the Earth–star frame aren't synchronized; in fact, the star clock already reads 16 years when Earth and ship pass. Later, as the star and ship pass, the ship clock has advanced from 0 to 15 years. The star clock, running slow from the viewpoint of the ship frame, has advanced only 9 years. But since it was ahead to begin with, it now reads 25 years. The interpretation of the clock readings in the ship frame is that 15 years have elapsed in that frame while only 9 years have elapsed in the Earth–star frame.

Yet observers in both frames agree about what the clocks actually read whenever two clocks are right next to each other so observers can unambiguously compare times. In particular, observers in both frames agree that 15 years elapse on the ship's one clock. They also agree that the reading of the Earth clock at the Earth/ship passing differs by 25 years from the reading of the star clock at the star/ship passing. What they disagree about is the interpretation of this 25-year interval. To observers in the Earth–star frame the clocks at Earth and star are synchronized, so 25 years is a legitimately measured time between two events. To observers on the ship, Earth–star clocks are running slow, and advance only 9 years between the two events. The 25-year difference in clock readings occurs because the star clock is ahead of the Earth clock by 16 years.

Which interpretation is right? By now you should have enough faith in relativity to know that's a meaningless question. Observers in both frames are correct in asserting that the other's clocks run slow, and they're saved from outright contradiction by the fact that events simultaneous in one reference frame aren't simultaneous in another frame in motion relative to the first.

It's easy to conjure up situations in relativity that seem contradictory, and books on relativity are often full of paradoxes that seem to embody such contradictions. But there's always a way out, and it usually involves recognizing that simultaneity is relative. Careful consideration of the timing of events in two different reference frames almost always resolves any apparent contradiction.

PAST, PRESENT, FUTURE, AND . . . ELSEWHERE

● ● ●

That simultaneity is relative has just got us out of the seeming contradiction of observers in relative motion each finding that the
other's clocks "run slow." But in demonstrating the relativity of
simultaneity, I introduced what may seem an even more disturbing
thought—that the *time order* of events may depend on one's frame
of reference. Because relativity gives every uniformly moving reference frame equal status, this reversal of time order isn't just some
illusion. It's really true that I can observe event A to occur before B,
that you can observe B before A, and that we're both right. But how
can that be? Doesn't it wreak havoc with causality?

The Past Is History

Ask a historian what the past is and you'll get an answer something
like "all those events that have already occurred." What does it
mean that an event has already occurred? It means that the time of
that event is earlier than the time of the present event. But we've just
seen that the time order of events may depend on one's reference
frame, so how can the notion of the past have any meaning at all?
Or for that matter, the future?

There are, in fact, events that are unambiguously in the past—
meaning that they occurred at a time before the present event. Note
that I didn't say "before the present." I said "before the present *event.*"

That's because the relativity of simultaneity precludes my talking about a universal present instant that pervades the whole Universe. Back to the star-trip example: For an observer on Earth at the instant the ship passes, the present includes the events of Earth's clock reading 0 and the star's clock reading 0. But to an observer in the spaceship at the same instant, Earth's clock reading 0 is indeed an event in the present, but the star's clock reading 0 is not—since that event occurred 16 minutes earlier! So there's no such thing as a universal "present." There is, for me, the present event—namely, whatever is occurring *here* and *now*. "Now" isn't enough; I have to indicate "here" as well—and that means I'm talking about an *event*, not just a *time*.

What are some events that are truly in the past, meaning they unambiguously occurred before your present event, that is, the event of your reading these words? For one, your birth. There are no observers, in any state of motion, who would judge that event to occur after your here and now (although different observers will disagree about the *amount* of time between those events). We don't have to restrict ourselves to events in relation to the here and now. We can also ask, for example, whether the event of the *Titanic* hitting the iceberg preceded the event of the great ship's sinking. The answer is an unambiguous yes. Again, one event is clearly in the other's past. Consider also that in 1987 astronomers observed a supernova—an exploding star—in a neighbor galaxy some 160,000 light-years away. Clearly the supernova event itself occurred before the astronomers observed it, since it took light from the supernova 160,000 years to reach the astronomers' telescopes.

What do the three pairs of events we've just considered have in common? They're all causally related. Your birth is a necessary cause of your reading these words. Had the first event not occurred, the second could not have occurred either. Had the Titanic not hit the iceberg, it would not have sunk. Had the supernova explosion not occurred, the astronomers would not have observed it. In each case, the earlier event was capable of influencing the later one and, in fact, did influence it. That provides a more robust definition of the past: The past of a given event consists of all those events that are capable of influencing the given event. Similarly, the future of the given event consists of all those events that the given event can influence. Note again that I'm talking about past and future in relation to a specific event; in a Universe in which simultaneity is relative, there's simply no such thing as a universal past and a universal future. But when one event is in another's past, that relationship is not ambiguous. All observers will agree about which event came first (although, again, they may disagree on the amount of time between the events). So relativity doesn't violate causality, in that those events that are causally related have an absolute time order that no observer will dispute.

But are all pairs of events causally related? To get at that, I'll ask some questions that seem, at first, very strange: Are there events that have already occurred that are not in the past? Are there events that haven't occurred yet that are nevertheless not in the future? With our traditional definitions of past and future, the answer to both questions is obviously no, but with our "can influence" definition, the answer is less obvious. Are there events that have already occurred but that cannot have any influence on my here and now? Are there events that have not yet occurred but that cannot be influenced by what I'm doing right here and now?

Excitement on Mars

When NASA's Mars Rover was exploring the Martian surface in 1997, Mars was 10 light-minutes from Earth. That means it took light, as well as the radio waves used to communicate with Rover, 10 minutes to get from Earth to Mars or from Mars to Earth. You've probably heard that the speed of light is the maximum possible speed in the Universe—a consequence of relativity that I'll explore in detail in the next chapter. For now, let's accept this cosmic speed limit which, more accurately, states that information cannot be communicated at speeds faster than c. So that 10-minute travel time for light between Earth and Mars is the shortest time that any information could take to travel between the two planets.

Now, let's put ourselves in NASA's Mars Rover control room, during the time of the Rover mission. To make things concrete, let's say we're just beginning a coffee break. Suppose further that, 5 minutes before our break began, Rover's TV camera spotted a real, live

Martian strolling across its field of view. Is that event in our past? It happened 5 minutes ago, so it occurred before our present event. But could it influence us right now? Can we possibly know, right now, of Rover's discovery? No, and we won't be able to know for another 5 minutes, when we receive Rover's TV picture. So Rover's discovery of extraterrestrial life cannot possibly influence us, right now at the start of our coffee break here in NASA's control room. We, the Rover scientists, aren't excited by what's happened, because we simply can't know about it. By our "influence" definition, then, Rover's discovery is not in our past. It cannot be a cause of the event, namely our coffee break, that's occurring right now in the earthbound Rover control room.

We have in Rover's discovery on Mars and our earthly coffee break 5 minutes later a pair of events that just cannot be causally related, because no information can travel between them. They're too far apart in space and too close in time for even light to be present at both events. It's precisely such pairs of causally unrelated events that different observers, moving in this case relative to our Solar System, may see with different time orders. For example, there could be an observer for whom the two events are simultaneous and others for whom the coffee break occurs first. Still other observers would agree with us on Earth that the Martian discovery occurred first, but they wouldn't agree that the time interval was 5 minutes. I won't go into the mathematics of all this. But imagine in Figure 10.5 replacing Earth with Mars and the star with Earth. Then you can see that an observer moving in the direction from Mars toward Earth will observe Earth time advanced relative to Mars time. That advancement will be 5 minutes—meaning the two events will be simultaneous—if the observer happens to be moving at half the speed of light. Observers moving faster in the same direction will judge the coffee break on Earth to occur before Rover's discovery of Martian life. Observers moving slower but in the same direction will judge the Mars event to occur first, but by less than 5 minutes.

Of course all these observers are correct, because they're all in uniformly moving reference frames that are equally valid situations for applying the laws of physics. As you well know by now, measures of time simply aren't absolute, but depend on one's reference frame. What's new here is that the *time order* of events may also depend on reference frame. But there's no contradiction. That's because the only events for which different observers claim different time orders are those that can't be causally related, because they're too far apart in space and too close in time for any influence traveling at *c* or less to get between them. Too far apart by whose measures? By anyone's. Even though the different observers disagree about the time between the events, and for that matter about the distance between Earth and Mars, all agree that the distance is greater than the distance light could travel in the time between the events. More on this in Chapter 13.

Poor Mars Rover! It's made a great discovery, but, alas, its navigation system has failed and it's heading straight toward a deep crater. We, in the NASA control room, realize this just as we start our coffee break, and we calculate that in 5 minutes Rover will topple into the crater and be destroyed. Is Rover's demise in our future? Can we send a radio signal to stop Rover and prevent the impending disaster? No; Rover is 10 light-minutes away, and we have only 5 minutes until Rover reaches the crater. Nothing we do at our present moment can influence what happens on Mars 5 minutes from now. So, according to our influence definition of the future, the event of Rover falling into the crater is not in our future. Here's another pair of events—the start of our coffee break and Rover's unfortunate accident—that cannot be causally related. And, sure enough, there could be other observers, moving relative to Earth and Mars, who find these events simultaneous, and still others for whom Rover's demise occurs before our coffee break starts.

The Elsewhere

If the events of Rover's alien discovery and its crater accident aren't in our past or our future, where are they? Relativity opens a new realm of time—or, more precisely, of *spacetime*. Events that aren't in the past or the future of a given event are in its *elsewhere*. They're events that cannot communicate with the given event, so the two cannot be causally related—and different observers can disagree about their time order. On Mars, for example, any events occurring between 10 minutes before and 10 minutes after the start of our NASA coffee break are in the elsewhere of the coffee-break event. Similarly, events in the 2-million-light-year distant Andromeda galaxy are in our elsewhere if they occurred anytime more recently than 2 million years ago or will occur within the next 2 million years. Even events that occur on the other side of the room you're sitting in are in your elsewhere, provided they occurred less than a few nanoseconds (billionths of a second) ago or will occur less than a few nanoseconds from now. Clearly, the farther away a place is, the broader the range of time that lies in the elsewhere of your here and now.

You can picture the elsewhere with the help of Figure 11.1. Here I assume for simplicity that we live in a space with just one dimension—meaning we can move back and forth along a line but not in any other direction (it gets too hard or even impossible to draw with more dimensions!). I'll represent place—that is, where an event occurs—by its position along a horizontal line. I'll represent time—

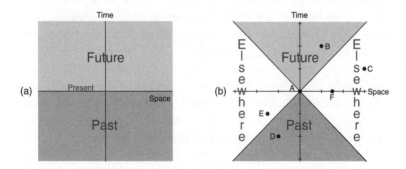

Fig. 11.1 Spacetime diagrams, showing only one spatial dimension. Each point in the diagram represents an event, its location in space given by its horizontal position, time by its vertical position. (a) In the Newtonian view, the present is the horizontal line separating future from past. All events happening right now, no matter where they are, occur in the present. (b) In relativity, future and past have different meanings for each event. Diagram shows the past, future, and elsewhere for event A. Tick marks are at intervals of 1 light-year in space and 1 year in time. The past comprises all those events that can influence A; the future those that A can influence. The text discusses the relations among the six events A–F.

when an event occurs—by its position along a vertical line. To make things simple, I'll measure position in light-years and time in years. Every point on the diagram represents a specific place and a specific time so the diagram is therefore a *spacetime diagram*. But time and place are what specifies an *event*, so each point on the spacetime diagram marks an event. The point at the center, where the time and space axes cross, is the here (position 0) and now (time 0).

So far, there's no relativity in all this, just a way of locating events on a simple diagram. Before relativity, people believed in a universal present that was the same instant for all, no matter where or what their state of motion. Figure 11.1a shows that situation. Any event at the present time lies on the horizontal line marked "present." Events below this line are at earlier times, or, according to our prerelativity definition, in the past. Events above the line occur at later times and are therefore in the future.

But relativity precludes information transmission faster than the speed of light and, therefore, puts a whole class of events into the elsewhere of the present event. Figure 11.1b shows the relativistic situation. I've marked six events on the diagram. To help locate them, I've put tick marks at intervals of 1 light-year on the horizontal (space) axis and 1 year on the vertical (time) axis. Event A is the here and now, or the present event (time 0, place 0). Event B occurs 2 years from now, and 1 light-year to the right of the present event. Since it's only 1 light-year away in space, there's plenty of time for light (or something slower, like a spaceship moving at half the speed of light) to communicate information from event A to event B. That is, A can influence B, and so B is unambiguously in A's future. What about event C? It occurs 3 light-years to the right of A, but only 1 year away in time. There's no chance for light or anything else to get from event A to event C. Thus A cannot influence C, so C is not in A's future but in its elsewhere.

Figure 11.1b also shows a pair of 45-degree lines through point A. Any event above A and between these lines is closer to A in distance (as measured in light-years) than it is in time (as measured in years); therefore A can influence such an event. All the events above A that are within the cone defined by these lines are in A's future.

Similarly, an event like D that occurred 2 years before A and 1 light-year away is clearly in A's past, along with all other events lying below A and within the cone defined by the 45-degree lines. Finally, events like C, E, and F that occur farther to the right or left than they do above or below A are in A's elsewhere. That C occurs after A, E before A, and F simultaneously with A is of no consequence because these pairs of events can't be causally related.

I've drawn Figure 11.1b from the viewpoint of a particular reference frame. Observers moving relative to this frame would have different measures of space and time, and their diagrams would be different. All observers would have events B and D in the future and past, respectively, of A, although how far away they are in time and in space would be different. All observers would have C, E, and F in A's elsewhere, but some would have C below the horizontal axis and some would have E above it. That is, for some observers event C would have occurred before A, and for some it wouldn't have. Still others would have A and C occurring simultaneously; others, A and E. And for some, F would occur after A; for others, before. But all that's OK because C, E, and F are in A's elsewhere and can't be causally related to A.

Note that I've been careful to say "the future of event A" or "in event A's elsewhere" but not "in the future" or "in the elsewhere." That's because the terms "past," "future," and "elsewhere" are meaningful not in relation to a time alone, but in relation to an *event*. In Figure 11.1b, for example, event B is in event A's future, but it's not in event C's future because it's 2 light-years to the left but only 1 light-year later in time. Drawing diagonal lines through C rather than through A would make this clear. Similarly, event E is in A's elsewhere but in D's future, while F is clearly in B's past.

You might wonder why the concept of the elsewhere isn't more obvious, given the large region it occupies in Figure 11.1b. Again, the answer lies in our limited experience. We don't tend to move around at high speeds relative to our surroundings or to deal with objects so distant that the travel time for light is significant. And we measure distances in, say, miles, and times in hours. In those units the diagonal lines in Figure 11.1b would not be at 45 degrees, but at a very low angle, hardly distinguishable from the horizontal axis.

That's because those lines represent the speed of light, which is a great many miles of distance for each hour of time. So the picture in our everyday units of miles and hours would look much more like the Newtonian view of Figure 11.1a. Only at great distances to the right or left of the here and now point would we notice that the very slightly sloping speed-of-light lines left a small wedge of elsewhere between past and future.

The elsewhere may sound like some mysterious new realm that's forever inaccessible to you. But it's not; for you, the elsewhere comprises those events that can't influence or be influenced by you as you are *here* and *now*. An event 5 minutes ago on Mars is in your elsewhere if you're here on Earth, but 5 minutes from now it will be in your past and at that time you can know of the Martian event and be influenced by it. Similarly, you can't do anything to influence an event that's going to occur 5 minutes from now on Mars, so it's in the elsewhere of your here and now. But if you had known an hour ago (or any time more than 5 minutes ago), you could have taken steps to prevent or alter the Martian event. That is, it was at one time in your future. But now it's in your elsewhere, and 15 minutes from now it will be in your past. Its relation to *you* changes because you're not an *event*. Its relation to a fixed instant in your life doesn't change, because that relation is truly between two events.

If you're not an event, then what are you? On a spacetime dia-

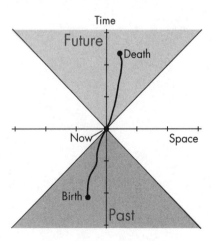

Fig. 11.2 The worldline of a human life is the continuous sequence of events between birth and death. This individual was born some distance to the left of where she is now and is just under halfway through her life. Can you find a period when the individual stayed pretty much in the same place? Note that the worldline is never at more than 45 degrees to the vertical, meaning that all events in the individual's life are causally related to the life events that precede them.

gram, you're a path, called a *worldline*. Each moment in your life occurs at some time and some place, so each moment represents an event, or a point in a spacetime diagram. Since you can't jump abruptly to a different time or a different place, the events in your life form a continuous sequence of spacetime points. The sequence advances inexorably into the future, but you're free to move about in space (in reality, you can move in all three dimensions, but only left or right in my simplified spacetime diagram). Figure 11.2 shows a spacetime diagram with the worldline of a human life. Note that no part of the worldline path can be inclined at 45 degrees or more to the vertical in a diagram scaled like Figures 11.1b and 11.2, since that would imply travel at light speed or faster. Put differently, each point on an object's worldline must be in the future of the points preceding it. What about a light beam? Its worldline is a 45-degree line, showing that light advances 1 light-year in space for each year of time. In more advanced treatments of relativity, worldlines play a crucial role in describing occurrences ranging from our star-trip example to the plunge into a black hole.

FASTER THAN LIGHT?

• • •

The premise behind the past-present-future-elsewhere structure of spacetime that I introduced in the previous chapter is that information cannot be transmitted faster than the speed of light, c. Were that premise wrong—in particular, if the instantaneous transmission of information were possible—then past and future would be unambiguous and there would be no elsewhere. Furthermore, with instantaneous transmission of information, we could synchronize all clocks everywhere with a single instantaneous signal, and there would be a universal time for everyone after all. Finally, events that we thought could not be causally related actually could be. Yet relativity tells us that those events can have their time orders reversed, which means that for some observers effect would precede cause! You can see that faster-than-light communication would wreak havoc with causality. Fortunately, it's not possible. Why not?

First Principles

First I'm going to give a quick and easy answer, one that follows directly from the Principle of Relativity. It won't be particularly satisfying, but it will be correct. In the sense that all the implications of relativity follow from the principle, it will be as complete as can be. Later, though, I'll provide answers you may find more satisfying.

The relativity principle states that the laws of physics are the same in any uniformly moving frame of reference. Among those laws are Maxwell's equations of electromagnetism, and among the predictions of Maxwell's equations is the existence of light waves going at speed c. That prediction must hold in all reference frames, meaning that all observers must measure the same speed c for light. I've used that argument many times before, so it should be tediously familiar.

Now suppose you and I are standing together as a light beam goes by. You hop into a fast rocket and try to catch up with the light. If you succeed, then you'll be moving with speed c relative to me, and you'll be at rest with respect to the light. If you were at rest with respect to the light, then you would be measuring a speed for light, namely zero, which is not equal to c. But light is supposed to have speed c with respect to *any* uniformly moving reference frame. If it didn't, the Principle of Relativity would be violated. So your situation—moving with speed c relative to me and therefore being at rest with respect to the light—must be impossible.

This question of catching up with light puzzled Einstein from the age of 16 until he resolved it with his special theory of relativity. He imagined running alongside a light wave, so the wave would be at rest with respect to him. Einstein understood Maxwell's electromagnetism, and he knew that Maxwell's equations dictated a particular form for an electromagnetic wave, a form in which electric and magnetic fields are perpendicular to each other and to the direction of the wave's motion. What bothered Einstein was that a *stationary* structure of that form was not a valid solution of Maxwell's equations. In fact, the only valid solution had the electromagnetic wave moving at speed c. So there had to be something wrong with the idea of running alongside a light wave so that the wave would appear to stand still.

Before relativity, there was no real problem here. Maxwell's equations were thought to be valid in only one particular frame of reference, namely the frame at rest with respect to the ether. You wouldn't expect electromagnetic waves as seen from other frames to satisfy Maxwell's equations, and there would be no problem with an observer moving through the ether in such a way that an electromagnetic wave appeared at rest.

But take away the ether, as Einstein did, and require that the laws of physics be valid for all uniformly moving observers. In other words, insist on the Principle of Relativity. Then all observers must measure c for the speed of electromagnetic waves, and no observer can be at rest relative to such a wave. Thus the Principle of Relativity asserts that it's impossible for you to run alongside a light wave and see it at rest, so it's impossible for you to move relative to me at speed c.

This argument is logically airtight and follows directly from the relativity principle. But it isn't very satisfying, because it doesn't explain *why* you can't get yourself up to the speed of light. What actually prevents that? I'll now explore several more satisfying answers to that question. However, there's no new physical principle involved; like everything else in relativity, these answers, too, are ultimately grounded in the Principle of Relativity.

Leapfrogging to c

Here's an obvious way to beat the cosmic speed limit, c. Suppose we have a big rocket capable of going, relative to Earth, at three-quarters of the speed of light ($0.75c$). Inside the big rocket, we build a miniature version with the same technology. We fire up the big rocket and zoom away from Earth at $0.75c$. You climb into the small rocket, fire the engine, and soon you're moving at $0.75c$ relative to the big rocket (Figure 12.1). So now you must be moving at $1.5c$ relative to Earth, right?

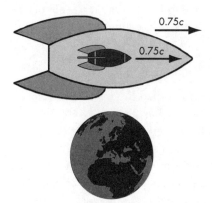

Fig. 12.1 The big rocket is moving past Earth at $0.75c$. A smaller rocket moves at $0.75c$ relative to the big one. So is the smaller rocket going at $1.5c$ relative to Earth? Common sense suggests it is, but in fact the speed of the small rocket relative to Earth is only $0.96c$.

Wrong! Why wrong? Because measures of space and time aren't the same in different frames of reference. The small rocket is indeed moving at three-quarters of light speed relative to the big rocket, and the big rocket is moving at three-quarters of light speed relative to Earth. But determining the speed of the small rocket relative to the big one involves measurements of distance and time in the reference frame of the big rocket. An observer on Earth doesn't agree with those measures and comes up with a different result for the speed of the small rocket. I won't go through all the math, but that speed is, in fact, 0.96c.

What I'm saying here is that 0.75c and 0.75c don't add to give 1.5c. That's an odd statement! Surely if I jog at 5 miles an hour down the aisle of an airplane going 600 miles per hour relative to the ground, then I'm going at 605 miles an hour relative to the ground. Why isn't it the same for the rockets? Actually, it is the same in both cases, but it's the example of the airplane that's wrong. In fact, my speed relative to Earth is a tiny bit less than 605 miles per hour—and for the same reason, namely that measures of time and space in the airplane's reference frame aren't quite the same as on Earth. Here the difference is negligible, because the relative speed of Earth and airplane is tiny compared with the speed of light. The effect becomes more dramatic as relative speed increases.

To get a little more abstract, suppose that the big rocket is moving at some speed u relative to Earth, and the small rocket is moving at speed v relative to the big one. Common sense suggests that the speed of the small rocket relative to Earth—call it v'—should be just $v' = u + v$. But relativity modifies this, giving

$$v' = \frac{u + v}{1 + u \times v}.$$

For obvious reasons, this equation is called the *relativistic velocity addition formula*. Here I'm assuming that all speeds are given as fractions of the speed of light. The numerator in the formula is just what we'd expect from common sense; it's the sum of the two speeds u and v. But the denominator has the effect of reducing that value. That effect is small if either of the speeds u or v is much less than the speed of light, since then the product $u \times v$ is much less than 1 and the denominator remains essentially 1. At high relative speeds, though, $u \times v$ is substantial and the speed of the small ship relative to Earth is a lot less than common sense would suggest. For example, let both u and v be 0.75 as in our example, and you'll see that the result is $v' = 0.96$.

It won't help to make the small rocket go even faster. No matter what its speed relative to the big rocket, as long as that speed is less than c then its speed relative to Earth will also be less than c. We can also ask what will happen if you shine a light beam inside the big rocket. Obviously, the light goes at c relative to the big rocket. What's its speed relative to Earth? Relativistic velocity addition tells us that, too. The light takes the place of the small rocket, so we put 1 in for v in our relativistic velocity addition formula (that's 1 because we're measuring all speeds as fractions of c, so the speed of light itself is just 1):

$$v' = \frac{u + 1}{1 + u \times 1} = \frac{u + 1}{1 + u} = 1.$$

In other words, the speed of the light is unchanged. It's the same relative to Earth as it is relative to the rocket, namely 100 percent of c. Of course, we already knew this; invariance of the speed of light is at the basis of relativity. Relativistic velocity addition, like all else, is based on the Principle of Relativity, so of course it gives a result consistent with that principle.

So we can't leapfrog to a speed greater than c by adding two sub-c motions. That rules out another idea you might have for achieving faster-than-light travel. Think of taking one of those people movers found in airports and put another people mover on top of it. People on the second mover should be going twice as fast relative to the ground as those on the first. Of course, that still isn't very fast. So pile more and more people movers, one on top of the other. Get enough of them and folks on the uppermost one should be going faster than c relative to the ground. You should be able to get to that superluminal speed by jumping from one mover to the next, undergoing just a tiny increase in speed each time. But it won't work. Even on the second mover, you aren't going quite twice as fast as people on the first one. That slight discrepancy compounds, according to relativistic velocity addition, so no matter how many people movers you pile up, none will be going faster than c relative to the ground.

Incidentally, understanding relativistic velocity addition helps dispel the common misconception that it's only light that exhibits unusual behavior. The unusual behavior of light is, of course, that it has the same speed for all observers regardless of their states of motion. For an object moving at less than c relative to one observer, another observer in motion relative to the first will measure a different speed for the object. This makes it seem that the unusual behavior—invariant speed—is unique to light. But as the relative speeds of observers and object approach c, the speeds measured by the two observers become nearly the same and their difference becomes much less than the relative speed of the two observers. So this relativistic velocity addition effect applies to everything, not just light. The only distinction is that with light there's no difference whatsoever in the speed as measured by different observers. For objects with relative speeds below c, there is a difference, but that difference becomes very small as the speed approaches c. Again, the effect here is not about light but about something more fundamental, namely, the nature of space and time.

$E = mc^2$

$E = mc^2$ is surely the most famous equation in all of physics, and for most people it's synonymous with Einstein and relativity. It's also, in the popular mind, the basis of nuclear weapons. So what's $E = mc^2$ doing here, a seeming footnote buried deep in a chapter on why faster-than-light travel is impossible?

In fact, $E = mc^2$ is itself a footnote to special relativity. It doesn't appear in Einstein's famous paper of June 1905. Later that year, Einstein published a second paper in which he asserted, "The mass of a body is a measure of its energy content," or, stated mathematically, $E = mc^2$. He elaborated more fully on this relation between mass and energy in a 1907 paper.

$E = mc^2$ is important but it's not the essence of relativity, and it's no more the basis of nuclear weapons than it is of a burning candle or a cave dweller's fire. Here I'm going to explore the meaning of $E = mc^2$, then come round to show how it gives what is probably the most satisfying reason that faster-than-light travel is impossible.

$E = mc^2$ asserts a fundamental equivalence, or, more precisely, an interchangeability, between matter and energy. Before Einstein, matter and energy were the two substances that populated the physical Universe. I use the word "substance" here in the sense of being "substantial"—that is, having ongoing, indestructible, existence. Matter is obviously substantial; it's the stuff we and everything else seem to be made of. Matter is quantified by its mass, a measure, roughly, of how much of it there is. We can transform matter from one form to another, as in chemical reactions. Burning coal is an example; carbon in the coal combines with oxygen in the air to make carbon dioxide. The total amount of matter seems unchanged; what's happened is a rearrangement of atoms to make a new chemical substance. But the atoms themselves seem unchanged. In other words, matter seems to be conserved; it can be rearranged, but not created nor destroyed.

Energy is less tangible than matter, but it too was thought in pre-relativity times to be conserved. Like matter, energy can change forms. Sunlight falls on a dark surface, transforming what was the energy of electromagnetic waves (light) into the random molecular motions we call, loosely, heat. Falling water turns an electric generator, transforming the energy of motion into electrical energy. Coal burns, transforming chemical energy into heat. You charge a battery, turning electrical energy into stored chemical energy. You put the battery in your laptop computer, and the chemical energy changes back into electrical energy to run the computer. You step on your car's brakes, changing the energy of the car's motion into useless heat energy (unless you've got a hybrid or electric car, in which case the energy goes back into the battery for later use). I could go on and on, since transformations among different forms of energy are involved in virtually everything that happens. In all these transformations, it seems that the total amount of energy remains unchanged.

What $E = mc^2$ says is that matter and energy are interchangeable. A piece of matter with mass m could, in principle, be transformed into pure energy, where the amount of energy, E, is the product of the mass m and the square of the speed of light. Because c is so big, this means that a little bit of matter could yield a large amount of energy. The equation works the other way, too. It says that an amount of energy E could be transformed into matter with mass m given by E/c^2. So matter and energy are, individually, no longer conserved. What is conserved is a new universal substance, which for lack of a better name we might as well call *mass-energy*. The total amount of mass-energy remains the same, but how much of it is in the form of mass and how much of it is energy can change.

Is the conversion of matter to energy and vice versa really possible? Yes, but total conversion occurs only in very special cases. It turns out that every one of the elementary particles that makes up everyday matter has an associated *antiparticle*, identical in mass but opposite in electric charge and other properties. Corresponding to the negatively charged electron, there's a positively charged anti-electron, also called a *positron*. Corresponding to the positively charged proton is a negative antiproton. Collectively, these antiparticles constitute *antimatter*. Our Universe, for reasons that are not yet entirely clear, seems to consist almost entirely of matter. A few antimatter particles are created in high-energy collisions and in some nuclear reactions, either naturally or in laboratory experiments. When a matter particle and its antimatter opposite meet, the two annihilate, disappearing altogether in a burst of energy. If the particle and antiparticle each have mass m, then the total mass that disappears is $2m$, and the energy that appears in its place is therefore $2mc^2$. The opposite process can occur, too. Under the right conditions, energy, in the form of a pulse of electromagnetic waves, can disappear and in its place a particle and its antiparticle come into existence. Figure 12.2 shows such a *pair-creation event*, as observed with a detector used in high-energy physics experiments.

Complete conversion of matter to energy is such a powerful process that if you had a box of ordinary raisins and a box of anti-raisins (much harder to get hold of!), the energy released in the annihilation of one raisin–antiraisin pair would be enough to supply all of New York City's energy needs for a day. If you had a power plant capable of harnessing that energy, you could drop in one raisin and

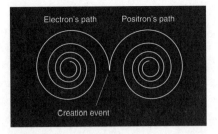

Fig. 12.2 Simulated image of a pair-creation event, as observed in the particle detector of a high-energy physics experiment. The detector records the paths of electrically charged particles. An uncharged, and therefore invisible, gamma ray—a high-energy bundle of electromagnetic energy—has entered the region from below. At the point marked, it ceases to exist and its energy becomes that of an electron and its antiparticle, a positron. A magnetic field points perpendicular to the page and causes the two new particles to spiral in opposite directions.

one antiraisin each day and that would be all the fuel you'd need to keep New York going.

Complete conversion of matter to energy is a rare event in the present Universe. (However, in the earliest instants of the Big Bang both it and the reverse were common and important processes.) But $E = mc^2$ isn't just about matter–antimatter annihilation and other exotic happenings. *Any*time a process results in release of energy, there's a corresponding reduction in mass. Weigh a candle, for example, and then light it. Weigh all the oxygen that sustains the flame. Capture all the smoke and carbon dioxide from the burning candle. Weigh those products, along with what's left of the candle, and you'll find that it all weighs slightly less than the original candle and the oxygen. How much less? Less, in mass, by the amount E/c^2, where E is the energy released (as heat and light) by the burning candle. Because c is so big, that's a very small amount of mass— so small that it would be virtually impossible to detect. Similarly, if you weigh a rubber band first when it's stretched and then after you release it, you'll find it weighs less. Again, how much less depends on how much energy you put into stretching the rubber band. Or weigh all the uranium fuel going into a nuclear power plant and weigh the fuel again after the plant has run for a year; you'll find there's less total mass. How much less? Less by an amount E/c^2, where E is the total energy extracted from the fuel, in the form of heat and electricity. The only difference between the nuclear power plant and the candle is that the nuclear plant converts a greater fraction of its mass into energy. $E = mc^2$ applies to both processes so to claim that it's necessary to know $E = mc^2$ to build nuclear weapons is to say that it was also necessary for our ancestors to know $E = mc^2$ before they first harnessed fire. Again: $E = mc^2$ applies to *all* processes that release or absorb energy.

Let me make this last point very clear with my nuclear power plant example. Suppose a nuclear power plant and a coal-burning power plant produce electricity at the same rate. Then in 1 year, each converts the same amount of matter into energy. The only difference is that nuclear processes are about 10 million times more effective in turning matter to energy than are chemical reactions. For that reason the coal-burning power plant has to process a lot more matter than the nuclear plant. A typical coal-burning plant consumes several 100-car trainloads of coal each week, while a nuclear plant might be refueled every year with a truckload of uranium. $E = mc^2$ applies to both power plants; it's just that the nuclear plant needs less fuel because it converts a greater fraction of its fuel mass to energy.

So why is Einstein often associated with nuclear weapons? First, because nuclear reactions are the most obvious and dramatic manifestation of mass-energy equivalence. In fact, it's only for nuclear reactions that reliable measurement of mass changes is possible. Measurement of those changes, in fact, gave physicists a new way of determining the speed of light: If you know the energy E released in a nuclear reaction and can measure the mass change m, then you can calculate c from $E = mc^2$. Einstein himself speculated as early as 1905 that newly discovered radioactive elements might contain so much energy that they could be used to test mass-energy equivalence, although 2 years later he ruled a nuclear test of $E = mc^2$ "out of the question." A second reason for Einstein's association with nuclear weapons is historical. In a famous letter written in the summer of 1939, Einstein informed President Franklin Roosevelt of the wartime potential of nuclear energy and urged the president to ensure that the Allies did not fall behind the Germans in nuclear weapons research. The ultimate result was the Manhattan project that developed the world's first nuclear weapons. But the Roosevelt letter was not Einstein's idea, and at the time Einstein was not particularly familiar with developments in nuclear physics. Leo Szilard, who had first conceived the idea of a nuclear chain reaction, visited Einstein to urge that the great physicist lend his name to the effort to mobilize nuclear weapons research. Einstein and Szilard jointly composed the letter, which Einstein alone signed, and a mutual acquaintance of Szilard and Roosevelt then delivered the message. So it was Einstein's fame, not the substance of relativity theory, that accounts for his only direct influence on the development of nuclear weapons.

Back to $E = mc^2$. I've emphasized that this expression of mass-energy equivalence applies to *all* forms of energy. That includes the energy of motion—what physicists call *kinetic energy*. The faster an object is going relative to you, the greater its kinetic energy. Because mass and energy are equivalent, the extra energy associated with an object's motion manifests itself in the same way that mass does—it makes it harder to change the object's motion. For example, a massive bowling ball is harder to get moving than a much lighter tennis ball. According to Einstein, it's even harder to get the ball moving faster if it's already moving relative to you. That's because its kinetic energy adds to the ball's inertia, or resistance to changes in its motion. In other words, both mass and energy have inertia.

So what? Imagine trying to get an object—for example, that bowling ball—up to the speed of light. At first it's not too hard to make it go faster, but as its speed approaches a substantial fraction of c, the inertia of its kinetic energy becomes substantial, and a given force becomes less effective at increasing the ball's speed. As the speed approaches c, Einstein showed, the force and energy required to make the ball go still faster both grow rapidly larger. As the ball's speed approaches c, the energy you supply in attempting to make it go even faster instead goes mostly into increasing its inertia. To get all the way to c would require infinite force and infinite energy. That's simply impossible, and for that reason you can't get a material object up to the speed of light. You may find this argument more satisfying than my earlier ones, because you can imagine pushing on that bowling ball and finding that it now has so much inertia that you just can't increase its speed very much. But again there's no new principle involved here; mass-energy equivalence, like the rest of relativity, follows ultimately from the Principle of Relativity.

Tachyons, Luxons, Tardyons

All I've really demonstrated is that it's not possible to get an object from rest (relative to you) up to the speed of light. Might there be objects that move always with speeds greater than light? Some physicists have explored that possibility, calling hypothetical faster-than-light particles *tachyons* (from the Greek "tachy" meaning "swift," as in your car's tachometer that measures engine speed or the pathologically fast heartbeat termed tachycardia). If tachyons exist, they can never be slowed down to speed *c* or below, and in particular we and they can never be at rest relative to each other. Experimental attempts to detect tachyons have failed, and many physicists believe that they might permit causality-violating time travel into the past. For that reason alone their possible existence is highly questionable.

It's also possible that things other than light could travel at speed *c*. Collectively, entities that move at light speed are called *luxons*. For decades physicists believed that elusive particles called neutrinos belonged to this class. Luxons must have zero mass, and for that reason their inertia doesn't become infinite at *c*. In fact, luxons can only move at *c* and cease to exist when stopped. Experiments in the late 1990s showed that neutrinos almost certainly have a very small mass, and if that's true then they can't move at the speed of light. Hypothetical particles called *gravitons*, associated with gravity, are predicted to go at *c*. For now the only certain luxons are the *photon*, which in quantum physics describes the particlelike bundles of energy that make up light and other electromagnetic waves, and the *gluon*, a particle involved with the strong force that binds atomic nuclei.

Tachyons, luxons, . . . what else might there be? Well, there's all the rest of us material entities with greater-than-zero mass. We're *tardyons*, from the root "tardy," for "slow" or "late." We're constrained to move only at speeds less than c relative to any uniformly moving reference frame, and it's to us that the "cosmic speed limit" arguments of the preceding sections apply.

Quantum Weirdness

It's not so much faster-than-light speed that relativity prohibits, rather, it's the transmission of information at speeds greater than c. As we saw in the previous chapter, that prohibition is what saves the logical order of cause and effect from the otherwise devastating fact that different observers ascribe different time orderings to the same pairs of events. So tachyons and other superluminal occurrences aren't strictly banned, but if they exist they have to do so in such a way that precludes information transmission at faster-than-light speed.

Recently there's been considerable interest in a strange phenomenon in quantum physics whereby two objects are somehow "entangled" in such a way that they seem able to communicate instantaneously even when they're far apart. This unusual possibility was first suggested in 1935 by Einstein and his colleagues Boris Podolsky and Nathan Rosen. Einstein, Podolsky, and Rosen thought up the so-called EPR paradox as a way of questioning the probabilistic interpretation of quantum physics—an interpretation that Einstein rejected, most famously in his claim, loosely translated, that "God does not play dice with the Universe." Much later, in 1964, the British physicist John Bell showed how experiments on the EPR phenomenon could distinguish between probabilistic and deterministic interpretations of quantum physics. Experiments done in the 1980s and subsequently have confirmed the probabilistic interpretation, upholding the apparently instantaneous communication between two widely separated particles.

In a typical EPR experiment a pair of particles is created and moves away from each other in opposite directions. The particles have certain properties that can take on particular values; for one property those values are described by the words "up" and "down." It's known for certain that if one particle is up, then the other must be down. Now here's the weird thing: according to the probabilistic interpretation of quantum physics, the particles don't have well-defined values "up" or "down" until someone measures them. It's not just that those values aren't known before measurement—it's that they're not even *determined*. So the particles fly apart and, later, experimenters measure one of the particles to have the value "up." According to quantum physics, the act of measurement is what causes the particle to have that value; before the measurement it was undetermined. When the one particle is measured to be up, then the other must, immediately, become down. Indeed, experiments confirm that this is the case. Somehow the act of measuring the up or down state of one particle in the pair has instantaneously fixed the state of the other—even though it's far away.

Is this superluminal communication? Are the two particles in violation of special relativity? Physicists think not. The two particles have, in some weird quantum-physics sense, a correlated existence. Even when separated, they act somehow like a single entity. Perhaps this entangled pair of particles does communicate instantaneously. Here's the rub: there's no way you can use this strange entanglement to send information, because it's entirely random whether you measure up or down for the first particle. So you can't say, to paraphrase Paul Revere, "'Up' if by land and 'down' if by sea . . . ," because you have no way of influencing whether your measurement of the first particle will yield up or down.

The EPR phenomenon is philosophically fascinating and reveals a startling interconnectedness at the heart of physics. You can find out a lot more about EPR from works on quantum physics. Here, in a book on relativity, it's enough to note the phenomenon and to recognize that it appears not to violate relativity's prohibition on transmitting information at speeds greater than c.

Reports appear from time to time on other seemingly superluminal phenomena, often in astrophysical situations. So far, all have proven to have other explanations. So solidly grounded is special relativity that if you hear of a new discovery of something moving faster than light, be skeptical! But, like any good scientist, remain open-minded too. By the way, I need to assert again that the cosmic speed limit is the speed of light in vacuum, c, and that it applies to the transmission of information. It's entirely possible for an object to move through a material medium at a speed faster than the speed of light *in that medium*. For example, the speed of light in water is about two-thirds of c. When electrons or other particles move through water at faster than this speed, they produce shock waves analogous to the sonic booms from supersonic aircraft. These shock waves are essentially intense emissions of electromagnetic radiation, produced when waves pile up on each other because they can't move fast enough to get away from the rapidly moving particles. Some very high energy sources of x-rays and visible light, used in research and technology, take advantage of this mechanism. You may also hear of certain kinds of waves with speeds purportedly greater than c. In the mathematical description of wave motion, there are two different speeds associated with waves. One of them can, indeed, be greater than c—but this speed is not associated with the transmission of any information. The second speed, that of the information carried in the wave, is always less than or (for electromagnetic waves in vacuum) equal to c.

The cosmic speed limit c really does appear to be inviolable!

IS EVERYTHING RELATIVE?

• • •

It's been fashionable to invoke Einstein in support of the idea that everything is relative not only in physics but also in distinctly non-physical realms like morality, aesthetics, and politics. I'm not going to get into those philosophical arguments here, but I do want to look closely at whether "everything is relative" is true in physics.

Physics is the science that tells us how physical reality works. What we might expect of physics is a truly objective description of that reality, a description that transcends one's own narrow point of view. Physics shouldn't be so much about those things that vary from one viewpoint (read "frame of reference") to another, but about underlying, objective truths that everyone agrees on. That may sound lofty and grandiose, but it highlights the reason why a physics that offered nothing more solid than shifting measures of time and space couldn't be altogether satisfying.

In fact, everything isn't relative. Relativity theory itself offers a permanence, an objectivity, that transcends the stretching of time and squeezing of space that we find justifiably fascinating. Einstein himself was not thrilled with emphasizing primarily what's relative. He did not call his own work the theory of relativity and accepted that terminology only after it had become widely established. Further, Einstein sympathized with a group of physicists who, in the 1920s, sought to change the name to "theory of invariance." But "theory of relativity" stuck, even though it describes a theory in which, in fact, everything is not relative.

So what isn't relative? If such seeming absolutes as space and time turn out to be relative to one's frame of reference, what can be left that's truly absolute? The Principle of Relativity contains one answer: the laws of physics are absolute, applying equally well in all uniformly moving frames of reference. A corollary that I've stressed repeatedly gives us another absolute: the speed of light, c, has the same measured value for all observers in uniformly moving reference frames. Are there other absolutes, quantities whose values don't depend on one's frame of reference? Yes, there are many.

Spacetime

The eminent relativity physicists Edwin Taylor and John Archibald Wheeler begin their delightful book *Spacetime Physics* with the "Parable of the Surveyors" (check out the Further Readings list for details). Here they describe a fictitious town that is mapped by two different surveyors. One works in the daytime and uses a magnetic compass to establish directions. The other works at night and takes directions from the North Star. As anyone who has worked with map and compass knows, magnetic and true north are not quite the same; hence, the surveyors' maps are slightly different, even though they map the same physical town.

Suppose you want to get from point A to point B in Taylor and Wheeler's fictitious town. You check the daytime surveyor's map (Figure 13.1a) and see that you can do this by going 4 miles east and then 3 miles north. But if you use the nighttimer's map (Figure 13.1b) you'll want to go nearly 5 miles east and then just short of 1.5 miles north. Which map is right? They're both right, of course. Either way, you get from A to B. It's just that one surveyor's definitions of north and east aren't the same as the other surveyor's definitions. Each has produced an accurate map, and if you follow either surveyor's directions carefully, you can get to any point you want.

The daytime surveyor says, "Point B is 4 miles east of A, and 3 miles north." The nighttimer says, "B is nearly 5 miles east of A, and just under 1.5 miles north." They disagree about how to describe the position of B relative to A, but there is something both agree on:

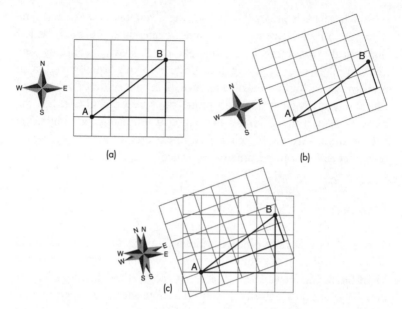

Fig. 13.1 (a), (b) Two maps describing the same region but using different definitions of the compass directions. Instructions for getting from A to B differ depending on which map one is using, but, as (c) makes clear, the straight-line distance from A to B doesn't depend on one's choice of map.

If you walk in a straight line from A to B, rather than following compass directions, you will walk a distance of 5 miles. Figure 13.1c shows this obvious fact. On either map, the straight-line distance is the hypotenuse of a right-angled triangle. Either surveyor can calculate the length of that hypotenuse using the Pythagorean theorem. Despite the fact that they're working with different triangles whose sides have different lengths, both get the same value—it happens to be 5 miles—for the straight-line distance between points A and B.

This is all very obvious, but let me make clear why it is so. It's because the distance between points A and B is an objective fact about the world, a fact that does not depend on any particular choice of compass directions. A hundred different surveyors, with a hundred different choices for the direction of north, would come up with a hundred different maps and a hundred different sets of instructions for getting from A to B. But all would agree that the straight-line distance is 5 miles. In other words, the distance from A to B is an *invariant*, a quantity that doesn't depend on one's particular point of view as embodied, in this case, in the choice of which direction to call north. Invariant quantities can claim to be objectively real in a way that quantities dependent on one's point of view cannot.

Taylor and Wheeler's surveyor parable is an analogy for relativity. In relativity, the points A and B are not points in space; they are events characterized by their locations in both space and time. Spatial distances in relativity are like east–west distances in the fictitious town. They depend on one's point of view—in the case of relativity, on one's state of uniform motion rather than one's choice of north. Time intervals are like north–south distances. They, too, depend on one's point of view, that is, on one's choice from among the infinitely many possible uniformly moving reference frames. The spatial distance between two events and the time between those events are not absolute but relative to one's frame of reference. This, of course, you know from earlier chapters.

But everything is not relative, and there are absolutes even in relativity theory. In Taylor and Wheeler's parable, the different surveyors could use their different east–west and north–south distances to calculate the same, invariant distance between points A and B. Similarly, in relativity, observers in different reference frames can use their different measures of spatial and temporal separation between two events to find an invariant "distance" not in space alone but in *spacetime*. All uniformly moving observers, no matter what their states of relative motion, will agree on the value of this *spacetime interval* between any two events. Because its value is independent of one's point of view (read "frame of reference"), that interval can claim a reality that relativity denies to measures of space alone or time alone.

The different surveyors in Taylor and Wheeler's town chose to break the two-dimensional space of their town into north-south-east-west in different ways, but the underlying reality of the town itself didn't depend on those arbitrary choices. Similarly, we break the underlying reality of spacetime into space and time in different ways, depending on our states of relative motion. It's not space and time that have absolute, invariant reality, but rather their combination in the single entity called spacetime. Where the surveyors' space was two-dimensional, spacetime must have *four* dimensions because there are three independent spatial dimensions and one of time.

The German mathematician Hermann Minkowski, whose classes Einstein had attended as a student, later studied relativity and was first to conceive the idea of four-dimensional spacetime and the invariance of the spacetime interval. Minkowski summarized his insight in an oft-quoted statement: "Henceforth space by itself, and time by itself, are doomed to fade away into mere shadows, and only a union of the two will preserve an independent reality."*

Given our map analogy, you might expect that Minkowski's spacetime interval follows from applying the Pythagorean theorem to the space and time intervals between two events. Actually, though, the mathematics is different. Where the Pythagorean theorem gives the square of a right triangle's hypotenuse as the sum of the squares of the two sides, the spacetime interval in relativity has its square given by the *difference* of the squares of the time and space intervals. That simple change from a positive to a negative sign tells us that spacetime does not obey the rules you learned back in tenth-grade geometry. Furthermore, it raises the possibility that the square of a spacetime interval can be positive, negative, or zero. That mathematical division into three classes corresponds to Chapter 11's classification of event pairs as being either causally related or not. For those events that are close enough in space and far enough apart in time that an observer can be present at both—and thus the events can be causally related— the square of the spacetime interval is positive. Such an interval is called *timelike* because an observer with just the right motion can be present at both events, so for that observer the events are separated in time only. The numerical value of the interval is just the time read on that observer's clock. In other reference frames the separation is a mix of space and time, but the time separation always dominates; hence the name "timelike." On the other hand, if the two events are so far apart in space and so close in time that they can't be causally related, then the square of the spacetime interval is negative and no observer can be

---

*Hermann Minkowski, quoted in E. F. Taylor and J. A. Wheeler, *Spacetime Physics*, 2d ed. (New York: W. H. Freeman, 1992), p. 15.

present at both events. For an observer in just the right reference frame, the two events will be simultaneous. For this observer their separation is purely spatial; hence, the interval is *spacelike*. Take the negative sign off its square, take the square root, and the numerical result is the distance between the events in the frame where they're simultaneous. Between the general cases of timelike and spacelike intervals is the *lightlike* interval, whose numerical value is zero even though the two events are located at different places and occur at different times. Events separated by lightlike intervals correspond to the emission and arrival of a light flash. We'll soon see how the unusual geometry suggested by relativity's modified Pythagorean theorem becomes central to Einstein's general theory of relativity.

The Meaning of c

There's one further complication, and in Taylor and Wheeler's parable it involves an unusual religious convention. In their fictitious society, north–south distances are considered sacred and are measured in a special unit—the mile. East–west distances are not sacred and are measured in ordinary meters. Different surveyors cannot consistently describe the straight-line distance between points A and B because that distance involves two different units for measuring spatial separation. Their religious convention keeps them from doing the obvious—converting all distances to the same unit, be it miles or meters. Then a brilliant and creative new surveyor arrives in town. This newcomer compares the different surveyors' maps, and—heresy!—converts all distances to the same unit. Henceforth all surveyors find that they agree on the straight-line distance between any two points. In other words, they discover an invariant aspect of reality that doesn't depend on their different points of view.

Einstein is the newcomer. What his relativity shows is that time and space, before considered separate quantities, are actually aspects of the same underlying physical reality. That reality is obscured by the convention of measuring time in one unit (seconds, for example) and space in another (meters). In reality, we should use the same units for both. To do so we need to know the conversion between seconds and meters—just as the surveyors needed to know the conversion between miles and meters to put their measurements in the same units. What is that conversion in the case of meters and seconds? Simply the speed of light, c. What's a meter of time? It's the time it takes light to cross a meter of space. What's a second of space? It's the distance light goes in 1 second of time. We can use either unit for both space and time, or any other distance or time unit we choose. In this book we haven't gone quite all the way to treating space and time with exactly the same units. Instead, we've been measuring time in years and distance in light-years. But the effect is the same; the conversion between time and space units is the speed of light, in this case 1 light-year per year. Had we dropped the "light-" we would have been measuring both distances and times in precisely the same unit, namely, the year.

Viewed in this sense, the speed of light isn't so much a speed as it is a conversion factor between units of space and time. It's pure coincidence that it has the value 299 792 458 meters per second; that number is just an artifact resulting from our arbitrary decisions to define the meter and second as we first did—definitions that were made long before anyone knew anything about relativity. If we embrace relativity fully, we should dispense with one of those units and choose a single unit for both space and time. In that more philo-sophically sensible system, the speed of light is just the number 1. Of course, measuring time and space in the same units isn't very prac-tical for beings who creep about their planet at speeds much less than c. A 60-mile-per-hour highway speed limit would have to read 1/10,000,000 in those units! But when we're discussing hypotheti-cal high-speed space trips, or for physicists working with elementary particles moving through their labs at relativistic speeds, it makes eminent sense to use units where $c = 1$. In those units space and time are implicitly aspects of the same underlying reality.

What Else Isn't Relative?

Are there other relativistic invariants besides the space-time inter-val? There are, and each combines previously unrelated quantities into a single four-dimensional entity. Here I'll describe just a few more of these invariants, which necessarily involve physics concepts I haven't discussed in this book. If you're unfamiliar with these concepts, be content to think of the spacetime interval as the most obvious of many quantities that are, in fact, the same for all observers.

If you've studied high-school physics, you learned about *momentum*—the product of an object's mass and velocity that Newton called its "quantity of motion." You also know about energy, a concept I discussed in the previous chapter. In Newtonian physics, momentum and energy are two distinct quantities. In relativity, they're aspects of a single four-dimensional quantity sometimes called *momenergy*. Energy is the time part of momenergy and momentum is the space part. Observers in different reference frames will measure different values for an object's momentum and for its energy, but when they combine them using the same modified Pythagorean theorem I introduced in connection with the spacetime interval, they get the same value. In suitable units, incidentally, that value is just the object's mass.

Yet another invariant is related to electricity. *Electric charge* is a fundamental property of matter and is an especially important concept in the study of static electricity. After learning about static electricity, one usually goes on to study *electric current*, the flow of electric charge that powers our electrical and electronic devices. Static and current electricity may seem to have little in common, but in fact electric charge is closely related to the time part of a single four-dimensional charge–current entity whose space part is related to the electric current. (It's actually the *density* or concentration of charge and current that constitutes the space and time parts of this so-called four-dimensional current.) If you're at rest with respect to an electric charge, all you see are the static electric effects of that charge. If you move relative to the charge, then it's moving relative to you, and you see an electric current. Different observers see different mixes of charge and current densities. If they combine the two according to the modified Pythagorean theorem, they get a value on which all observers agree. Once again, there's an underlying four-dimensional reality that's the same for all observers. In electricity, how that reality breaks into charge and current concentrations depends on the observer's frame of reference, but the underlying reality is the same for all.

While we're on electricity, I'll give you just one tantalizing hint of another remarkable insight to be gained from relativity; to get the full flavor of this one, you'll have to dig into a more substantial physics book.* Suppose I have some electric charges moving through a wire. For simplicity, let's assume positive charge is moving one way and negative charge the other, giving a net motion of charge that constitutes an electric current (that isn't quite how currents in real wires work, but it will make the explanation simpler). Assume further that the positive and negative charges are equally spaced. Now let's put another charge in the vicinity of the wire; we'll call it a test charge. If the test charge is at rest with respect to the wire, it "sees" the wire as containing identical lines of positive and negative charge, moving in opposite directions. According to relativity, the distance between those charges is contracted compared to what it would be if the test charge were at rest with respect to them instead of at rest with respect to the wire. But it's contracted the same amount for both positive and negative charges, since they're moving at the same speeds relative to the test charge (albeit in opposite directions). So as far as the "stationary" test charge is concerned, the wire contains equal densities of positive and negative charge, so it's electrically neutral. It therefore exerts no electric force on the test charge.

Now suppose that the test charge moves alongside the wire. Relative to this "moving" test charge, the lines of positive and negative charge are moving with different speeds. So length contraction affects each line of charge differently and as a result the test charge no longer sees equal densities of positive and negative charge. It "thinks" the wire is electrically charged, and correspondingly it feels an electric force. That force is what we usually think of as magnetism—the force resulting on a charge that *moves* in the vicinity of an electric current. Viewed this way, magnetism is a necessary

*See, for example, R. Wolfson and J. M. Pasachoff, *Physics for Scientists and Engineers*, 3d ed. (Addison Wesley Longman, 1999), pp. 1035–7; the sections just before this also explore relativistic invariants. A more thorough analysis is given in E. M. Purcell, *Electricity and Magnetism*, 2d ed. (New York: McGraw-Hill, 1985), chap. 5.

consequence of electricity in a world where the relativity principle applies.

Once again, it's relativity that unifies two seemingly separate phenomena, in this case electricity and magnetism. What one observer sees as a purely electric effect another will see as a combination of electric and magnetic effects. How electromagnetic phenomena break out into separate electric and magnetic aspects depends on one's frame of reference. In that sense electricity and magnetism separately, like space and time separately, have no claim to objective reality. Underlying the relative measures of space and time is a single, objective measure of spacetime that's the same for all observers. Similarly, all observers can describe the electric and magnetic phenomena they observe as resulting from a single, underlying electromagnetic entity. Different observers disagree about the extent to which phenomena are electric or magnetic, but each can understand his or her observations as particular viewpoints on an underlying, objective electromagnetic entity.

Is Everything Relative?

We now have a firm answer to the question that forms this chapter's title. Everything is *not* relative. If it were, there would be nothing absolute about physics and that science could hardly claim to be a description of an objective physical reality. Cherished absolutes of our common sense—in particular, space and time—have lost their absolute status and have become relative to our particular point of view (i.e., frame of reference). But they're not lost for good; they've simply merged into the greater, more encompassing absolute of four-dimensional spacetime. Space and time aren't the only players on this four-dimensional stage; many other physical quantities that once seemed distinct have merged into four-dimensional entities with an absoluteness that transcends particular frames of reference. Ultimately, relativity has enhanced, not diminished, our sense of an objective and absolute reality.

CHAPTER 14

A PROBLEM OF GRAVITY

• • •

What's so *special* about Einstein's *special* theory of relativity? Special here doesn't mean there's something wonderful or noteworthy about Einstein's theory. Instead, special means specialized, limited, restricted—to the special case of uniform motion. You've seen that the essence of special relativity lies in a single statement: The laws of physics are the same for all observers *in uniform motion*. But why this restriction? Why should it matter whether motion is uniform? Why not the same physics for everyone, period?

These questions provide a philosophical motivation for generalizing the special theory of relativity into a statement that drops the restriction to uniform motion. The result would be, and is, the *general theory of relativity*. Going from the special to the general theory proves more complicated—philosophically, physically, and mathematically—than you might expect. It took Einstein some 10 years, with many false starts and wrong turns, from his 1905 paper introducing special relativity to the completion of his general theory in its final form. Along the way the general theory became not just a generalization of special relativity but a whole new way of understanding gravity.

This chapter explores the development of general relativity and shows why it became a theory of gravity. In the last two chapters I'll go on to examine the implications of general relativity and its central role in modern astrophysics and cosmology. The mathematical complexity of general relativity means that I can't give you the kind of ironclad logical development that the preceding chapters provided for special relativity. Instead, I'll try to motivate your understanding of the key ideas in general relativity and relate them to just a few basic principles.

A Problem with Newton

Einstein knew that Newton's theory of gravity could not be consistent with special relativity. In Newton's "action at a distance" view of gravity that I introduced in Chapter 4, Earth somehow reaches out across empty space and pulls on the Moon, a quarter-million miles distant, exerting the force that keeps the Moon in its circular orbit. In Newton's view, that influence from Earth to Moon is instantaneous. The Moon knows right now about the presence of the distant Earth, as it is now. Suppose Earth were suddenly to disappear. According to the action-at-a-distance view, the Moon would immediately cease to feel Earth's gravity and would go off on the straight-line path that Newton says it should follow in the absence of a force. But that implies instantaneous transmission of information from Earth to Moon, which is inconsistent with relativity's assertion that information cannot be transmitted faster than the speed of light.

Another problem with Newtonian gravitation is that the gravitational force depends on the distance between massive objects. We've learned that "the distance between . . ." is not an absolute quantity; it differs from one reference frame to another. So in whose reference frame is that distance to be measured? And exactly when, especially if gravitating objects are in motion relative to one another?

Incidentally, you might wonder whether electromagnetism suffers the same problems. Is special relativity also inconsistent with electromagnetism, with its description of the forces between electric charges and between magnets, and of electric and magnetic fields? The answer is no. As you've seen, relativity grew directly out of questions about the nature of light as an electromagnetic wave. Special relativity developed as a theory fully consistent with Maxwell's equations of electromagnetism. What saves electromagnetism is the interaction between electric and magnetic fields that produces electromagnetic waves—waves that travel at the speed of light. Consider an electron (negative charge) experiencing the attractive force of a nearby proton (positive charge). In the field picture, the proton creates an electric field in the space around it, and the electron responds to the field in its immediate vicinity. Suppose the proton were suddenly to disappear. In electromagnetism, news of that disappearance moves outward as an electromagnetic wave, traveling at the speed of light. Ahead of the wave, the electric field is just as it was before. Behind it, there is no electric field. The region of no field expands outward, from the site of the vanished proton, at the speed of light. Eventually the electron learns of the proton's disappearance as the no-field region reaches it and it ceases to feel an electric force. Because the information gets to the electron at the speed of light— rather than instantaneously as in the case of Newtonian gravity— there is no contradiction with relativity. In the field picture, furthermore, a particle responds only to the electric or magnetic field in its immediate vicinity. That means there's no problem with distances and times measured in different reference frames. Electromagnetism is fully relativistic already and needs no modification. Newtonian gravity is not relativistic, so a new, relativistic theory of gravity is needed.

A Shaky Foundation

The foundation of special relativity is the principle that physics is the same for all observers in uniform motion. How do you know whether you're in uniform motion? At first glance that seems easy. In an airplane, for example, you can tell the difference between smooth air and turbulence. In the former the plane's motion is uniform; in the latter it's anything but uniform. You know the motion in smooth air is uniform because things around you, like the airline peanuts you've spread over your seatback tray, stay put. In other words, things in the plane obey Newton's first law, the law of inertia: if they're at rest, they remain at rest. You could play tennis on this uniformly moving plane and the tennis ball would behave as you expect—that is, according to Newton's laws. Turbulent air, in contrast, would send your peanuts flying all over the place, and no way could you have a satisfactory tennis match. So it looks like you can tell pretty easily whether you're in uniform motion.

We can make the criterion for deciding if you're in uniform motion more concise and scientific: You're in a reference frame in uniform motion if, in your reference frame, objects obey the law of inertia. That is, if an object at rest in your reference frame remains at rest without the need for any forces to hold it at rest, then your reference frame is in uniform motion. Another, and actually better, name for such a reference frame is an *inertial reference frame*, so called because of the link to the law of inertia. On the other hand, if you need to apply forces to keep objects at rest—as you would have to do to your airline peanuts when the plane encounters choppy air—then you know you're not in an inertial frame.

Or do you? How do you know for sure that there aren't forces acting on objects to keep them at rest? If you thought there weren't, but there really were, then you would mistakenly conclude that you were in an inertial reference frame. After all, the most fundamental forces, like electricity, magnetism, and gravity, are invisible. Having an electric force acting isn't as obvious as seeing your hand holding the peanuts to keep them at rest. Maybe—just maybe—your airline peanuts are electrically charged and there's some clever arrangement of electric field holding them in place on the tray despite what might, in fact, be nonuniform motion. Farfetched? Yes, but you just might be deceived.

For electric and magnetic forces, fortunately, there's a simple test. Not all objects are electrically charged. So find one that is and one that isn't. Watch how they behave. If they behave the same way, then there's no electric force acting. If they behave differently, then the difference must be caused by an electric force. The same idea works for magnetism: some objects are magnetic and some aren't. So simple experiments will tell you whether electric or magnetic forces are present.

Gravity is different. Gravity affects all objects. Furthermore, as Galileo purportedly showed by dropping different objects from the Tower of Pisa, gravity affects all objects *in the same way*. That is, all objects, regardless of how massive they are, undergo the same acceleration in response to gravity. So gravity could be deceiving you. That is, you might think objects remain at rest because you're in a uniformly moving reference frame with no forces acting, when really gravity is acting to keep them in place in a reference frame whose motion is not uniform. Or, you might think that objects are undergoing accelerated motion relative to you because your reference frame is accelerating when in fact it isn't but the objects are accelerating due to the force of gravity. (If all this sounds a bit obscure, be patient; I'll be giving concrete examples in the next section.)

But surely you can tell the difference between a reference frame that is "really" in uniform motion and one that "really" isn't. Be careful! I cautioned earlier about using the word "really" in cases where relativity renders it meaningless, as in "You're really moving and I'm really at rest." Statements like that are ultimately meaningless because no physical experiment can distinguish absolutely between being at rest and moving. Only relative motion is meaningful. That, of course, is what special relativity is all about. Now we're finding that there seems in principle to be no way of telling with absolute certainty whether you're in uniform motion. So statements like "I am in uniform motion" are as meaningless as the statement "I am moving." This presents a big problem for special relativity, which is a theory about the laws of physics being the same for observers in uniform motion. If you can't tell for sure that you're in uniform motion, how can you ever know if special relativity is valid for you?

So special relativity is on shaky ground and, again, the culprit is gravity. The problem of instantaneous information transmission in Newtonian gravity means that we need a new theory of gravity and a more general theory of relativity that encompasses that new gravity. Now we have the additional problem that gravity make us unable to tell for sure whether we're in uniform motion. For that reason, too, we need a relativity theory that doesn't just apply to uniform motion. Such a general theory must necessarily consider gravity.

The Principle of Equivalence

My arguments in the preceding section may seem a bit abstract and unconvincing. Now I'll make them more concrete, following reasoning that set Einstein on the path to general relativity.

Imagine you're in a small, closed room at rest on Earth (picture an elevator car). You're holding a ball, which you proceed to drop. What happens? The ball falls to the floor, and it does so with a constant acceleration because of the constant force of gravity acting on it. That is, the ball starts from rest and as it falls it goes faster and faster. On Earth, in fact, the strength of gravity is such that the ball gains roughly 10 meters per second of speed (about 32 feet per second) for every second that it falls. Your dropping the ball constitutes a simple physics experiment, and it has a simple and precise outcome: the ball accelerates downward, gaining 10 meters per second of speed with each passing second. When I talk quantitatively about speed and acceleration here I am, of course, talking about quantities measured relative to the reference frame of the room. Figure 14.1a shows this simple ball-drop experiment.

Now let's attach a rocket motor to the room and travel to a distant region of intergalactic space, so far from any stars, galaxies, or other objects that gravity is truly negligible. We'll set the rocket firing to give

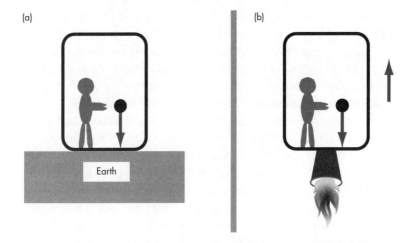

Fig. 14.1 The Principle of Equivalence. (a) A ball released in a reference frame at rest on Earth falls to the floor with constant acceleration. (b) Exactly the same thing happens in an accelerating reference frame in intergalactic space, far from any source of gravitation. Here the acceleration is upward and provided by the rocket motor. The situations of (a) and (b) are equivalent as far as the laws of physics are concerned.

an acceleration equal to 10 meters per second every second. That is, the force of the rocket motor is making the room go faster and faster, gaining 10 meters per second of speed in each second. You're standing in the little room, and again you perform the ball-drop experiment. What happens? While you're holding the ball, it shares your motion, and you in turn share the motion of the room. Why? Because the floor of the room pushes on your feet, exerting a force that gives you the same acceleration as the room. Your hand does the same to the ball. So the room, you, and the ball are all accelerating at the same rate.

Now, the instant you let go of the ball, there's no longer any force acting on it. So what does it do? It obeys the law of inertia, and it remains at rest relative to a uniformly moving reference frame that had the same speed as the room at the instant you released the ball. The room itself is not that uniformly moving frame, since it's accelerating. Relative to the uniformly moving reference frame I've just identified, the ball remains at rest and the floor of the room accelerates toward it at the rate of 10 meters per second in each second. What's this look like to you in the room? You share the room's motion, so it looks to you like the ball accelerates toward the floor, gaining speed (relative to you) at the rate of 10 meters per second in each second (Figure 14.1b). But this is just what happened when the room was at rest on Earth! From your point of view, inside the room, the outcome of the ball-drop experiment is the same in either situation. That experiment can't be used to decide whether you're in an unaccelerated reference frame in the presence of Earth's gravity or, instead, in an accelerating reference frame in the absence of gravity. Nor, in fact, can any other experiment involving the laws of motion.

In the accelerating room, incidentally, you also feel just as you would when standing on Earth. The floor of the room pushes up to accelerate you, and by Newton's third law you push down on the floor. The muscles in your body tense and compress just as they would when standing at rest on Earth. So your bodily sensations, too, give you no way to tell whether you're in an accelerating reference frame, absent gravity, or in an unaccelerated frame with gravity present.

Now consider two other situations. First, put the little room in intergalactic space again, but now turn off the rocket engine. What happens? Gravity is completely negligible, so when you release the ball, it just stays there. In this case the room is not accelerating, so it constitutes a uniformly moving reference frame in which the law of inertia is valid. The ball was at rest relative to this frame, and it remains at rest. It just floats there, next to your hand (Figure 14.2a).

Finally, consider that the little room is an elevator, whose cable has snapped. It's plummeting toward Earth, accelerating downward at that same 10 meters per second each second. You and the ball experience exactly the same acceleration, as shown by Galileo's purported experiment (and by many other more modern and very precise measurements). So when you release the ball, relative to you in the falling elevator room, it again just floats next to your hand (Figure 14.2b). In the two situations of Figure 14.2—being in a truly gravity-free environment and falling freely in the presence of gravity—the ball-drop experiment has exactly the same outcome. Within the

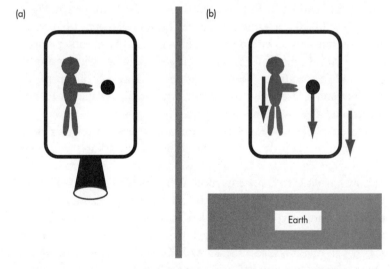

Fig. 14.2 Two other equivalent situations. (a) In intergalactic space, far from any source of gravity. With the rocket engine off, the reference frame is unaccelerated. You float about the room, and a ball released from rest doesn't go anywhere. (b) Exactly the same thing happens in a reference frame falling freely toward Earth because you, the room, and the ball are all accelerating downward at the same rate.

confines of the room, that experiment can't help you decide which situation you're in. Nor could any other experiment involving the laws of motion.

So the two situations of Figure 14.1 are indistinguishable, and so are the two situations of Figure 14.2. Indistinguishable, that is, by experiments involving the laws of motion. Why this indistinguishability? The reason, ultimately, is that all objects experience the same acceleration in the presence of gravity. In Figure 14.1b, the room's acceleration mimics the effect of gravity on any object, because all objects near Earth would fall with the same acceleration. In Figure 14.2b, the ball floats beside your hand because you, the ball, and the room all experience the same gravitational acceleration.

Because no experiment with the laws of motion can distinguish them, the two situations in Figure 14.1 are in a sense equivalent. So are the two in Figure 14.2. What is the ultimate reason for that equivalence? It's the fact that gravity has exactly the same effect on all objects. If that weren't true, different objects in Figure 14.1a would fall with different accelerations, but all objects in Figure 14.1b would still have the same acceleration relative to the room— and the situations wouldn't be equivalent. Absent other forces, objects with dramatically different masses really do fall with exactly the same acceleration. Apollo 15 astronaut David Scott dropped a hammer and a feather on the Moon; with no air resistance, they hit the ground at exactly the same time.

The simple fact of equal gravitational acceleration for all objects, known since Galileo's time, is nevertheless remarkably profound. Here's why: The mass of an object is a measure of its inertia, that is, of how hard it is to change its motion. That's what Newton's second law says; it tells us that it takes a bigger force to give a more massive object the same acceleration as a less massive object. There's nothing whatsoever about gravity in that statement. On the other hand, the mass of an object is also a measure of the force gravity exerts on that object. Drop a bowling ball and a tennis ball. It takes a much greater force to give the bowling ball the same acceleration as the tennis ball yet they fall with the same acceleration. Why? Because the gravitational force on the bowling ball is greater by just the amount needed to compensate for its greater resistance to being accelerated. In other words, mass as a measure of resistance to changes in motion goes hand-in-hand with mass as a measure of gravitational force. But why?

Why should an object's inertia—the property that determines its resistance to changes in motion—be essentially the same as the property that determines the force of gravity on the object? Until Einstein's time, this essential equivalence of two different meanings of the term "mass" was considered a coincidence, and physicists spoke of "inertial mass" and "gravitational mass" as distinct properties of an object that happened to have the same value. What Einstein saw, though, was a hint of a deeper relation between gravity and accelerated motion. Einstein elevated to the status of a fundamental principle the equivalence inherent in Figures 14.1 and 14.2, and the underlying equivalence between the two meanings of "mass." With the same incisively simplifying genius he showed when he applied the principle of special relativity to all of physics, Einstein declared the two situations of Figure 14.1, and the two of Figure 14.2, to be fundamentally indistinguishable by *any* physical experiments, not just those involving the laws of motion. This *Principle of Equivalence* is at the heart of general relativity.

Weightless!

You might object to my assertion of equivalence between the two situations in Figure 14.2 on the grounds that the hapless elevator occupant in Figure 14.2b will smash to smithereens when the elevator hits the ground. That's true, but beside the point. What I'm declaring equivalent are an unaccelerated reference frame in the true absence of gravity and a reference frame in accelerated motion *under the influence of gravity alone*. When the elevator hits the ground, strong nongravitational forces act on it and its unfortunate occupant, and the equivalence is no more. But there's a way around this unhappy outcome. There is nothing sacred about falling *down*. The equivalence would still hold if, in Figure 14.2b, I launched the elevator sideways, so it fell in a curving path. Because everything in the elevator would share its initial motion, all those objects would follow the same curving path and would appear to float relative to the elevator. Of course the elevator would still hit the ground, just not directly below its launch point. But—as Newton first suggested, and as modern spaceflight confirms daily—if I launched the elevator fast enough, and from a point above the atmosphere to avoid the force of air resistance, it would "fall" around the Earth in the circular path we call an orbit (recall Figure 3.4). Because all objects experience exactly the same acceleration due to gravity, you, the ball you attempt to drop, and everything else in this orbiting elevator would float freely about relative to the elevator. That's the origin of the apparent weightlessness that occurs in spaceflight. It isn't that there's no gravity in space or that objects in a spacecraft have somehow ceased to experience gravity. It's just that all objects experiencing only gravity have exactly the same acceleration, so they all "fall" together and thus experience no relative motion.

Being "weightless" has nothing, inherently, to do with being in space. It's just that in space one gets rid of the pesky, nongravitational force of air resistance and the danger of hitting the ground. NASA's "Vomit Comet" training aircraft executes an arcing flight that essentially cancels the forces of air resistance. Those onboard temporarily experience the same "weightlessness" as astronauts on the Space Station. "Weightless" scenes in the movie *Apollo 13* were filmed on the Vomit Comet. Reviews praised the film for its realism in "simulating the weightlessness of outer space." Nonsense! That was no simulation. *Any* reference frame under the influence of gravity alone is a reference frame in which objects are "weightless."

The state of motion under the influence of gravity alone is often called *free fall*: *free* because nothing other than gravity is acting; *fall* because gravity *is* acting to *change* the motion of an object in the direction toward a gravitating body like Earth or Sun. Most of us have trouble shaking the Aristotelian view that gravity should make things *move* downward, as opposed to the Newtonian view that gravity *changes* motion toward the downward direction without necessarily *moving* an object downward. For that reason free fall conjures up images of objects actually plummeting toward Earth. That state is, indeed, free fall—but so is the motion of the International Space Station, in its never-ending circular "fall" around Earth. For that reason the term *free float* is a more apt description of motion under the influence of gravity alone. The Space Station is in free float, along with everything inside it. So is the plummeting elevator, at least for a while. Free float simply means that the only influence acting is gravity.

Lost and Found: Uniform Motion

We've seen how confusion between gravity and accelerated motion makes it impossible to know for sure whether you're in a state of uniform motion. That puts special relativity—valid only for uniformly moving reference frames—on a shaky foundation.

To be specific, look again at Figure 14.2. Any experiment you do in the free-float situation of Figure 14.2b will give exactly the same result as it would in Figure 14.2a. We reasoned that conclusion for experiments involving motion, and Einstein's declaration of the Principle of Equivalence suggests it's true for an experiment involving *any* aspect of physics. So you might think you were truly in uniform motion in the absence of gravity (Figure 14.2a) when you were really in free float, accelerating downward in the presence of gravity (Figure 14.2b). You could be confused. That's the bad news.

Here's the good news: Because the two situations of Figure 14.2 are completely equivalent, all the laws of physics that work in the "truly" uniform motion situation of Figure 14.2a also work in the situation of Figure 14.2b. That includes special relativity. So the free-float reference frame of Figure 14.2b—or any other reference frame under the influence of gravity alone—is a reference frame where special relativity is valid! We've lost the ability to be absolutely sure whether we're in uniform motion, but it no longer matters because we've found a situation, namely free float, that is logically and physically equivalent to uniform motion.

The situation of Figure 14.2a is, in fact, impossible. Gravity is never completely gone, even in the emptiness of intergalactic space. But free float is easy. Jump off a chair and you're momentarily in free float. Board the International Space Station and you'll experience free float for as long as you stay. In free float, physics is simple. Objects obey the law of inertia, remaining at rest relative to the free-float reference frame unless some force acts on them. Some force, that is, other than the Newtonian gravitational force that acts with the same effect on everything in the free-float reference frame. Give an object a momentary push and after that it moves in a straight line at constant speed—relative, of course, to the free-float reference frame. Its path relative to Earth is more complicated, but that's beside the point. Physics is simple in the free-float reference frame.

Furthermore, physics works exactly the same way in every free-float frame. It doesn't matter whether that frame is the International Space Station, in its Earth orbit; a spacecraft on the way to Mars, as long as its engines aren't firing, so only gravity is acting; or a space probe orbiting the Sun or even plunging into a black hole. The laws of physics—all the laws of physics, according to Einstein and his equivalence principle—are just the same in every free-float reference frame.

So in the real Universe, with gravity, free-float frames take the place of the hypothetical uniformly moving frames that we used to build our understanding of special relativity. Since the law of inertia holds in free-float frames, I'll grace every one of them with the title "inertial reference frame." I won't call them "uniformly moving frames" because, at least on the face of it, they patently aren't. But in a real sense they're the closest we can get to uniformly moving frames, and they are every bit as good as "real" but impossible states of "truly" uniform motion. Well, almost as good (more on that shortly).

Because you can't tell for sure whether you're in uniform motion, the phrase "uniformly moving reference frame" is not particularly useful. Our newer term, intertial reference frame, is just fine. That term applies to a hypothetical reference frame in truly uniform motion, but it also applies to any free-float frame as well. In the real world, with gravity, the only inertial reference frames we have are free-float frames. The special theory of relativity applies in all inertial frames, and in the real world that means free-float frames.

Farewell to Gravity As We Know It

Stand on Earth and you feel the tug of gravity on your body. Or do you? Actually, what you feel are forces in your muscles as they tense or compress to keep your body from collapsing in a heap on the ground. You weary of the fight against gravity, so you frequently sit or lie down, letting chair cushions and mattresses support your body.

But for real relief, get yourself into a free-float reference frame. Board the Space Station or go skydiving (where you're essentially in free float until air resistance becomes significant and you approach a constant "terminal speed"). In free float, you don't feel gravity at all. That's why you're "weightless."

I've been putting "weightless" in quotes because, in the Newtonian view of things, you aren't weightless at all. Your "weight" means the force that gravity exerts on you, and that force is certainly present to keep you and your space station in orbit or to accelerate your skydiving body earthward. You don't feel that force because all parts of your body, and everything else in your free-float reference frame, experience exactly the same gravitational acceleration.

Now let's think about how Einstein would describe the situation. The essence of special relativity is that different inertial reference frames are equally good places to experience the laws governing the physical Universe. In relativity, something that has "reality" in one reference frame but not in another can't lay claim to truly objective reality. For example, the time between two events—an absolute, fixed quantity in Newton's view—depends, in fact, on which inertial reference frame you're in. So does the distance between two points or the length of an object. You saw all this in the earlier chapters on special relativity. In contrast, the spacetime interval I introduced in Chapter 13 can claim to be an objectively real quantity, since observers in different inertial reference frames find that it has the same value. Unless a quantity or phenomenon can garner this universal agreement among different reference frames, it can't claim to be "real."

The idea behind general relativity is to extend the Principle of Relativity to *all* reference frames, inertial or not. In general relativity, a phenomenon should be real only if observers in all reference frames agree about it. A phenomenon that's present in one reference frame but not another can't be objectively real.

That's how it is with what you now call gravity. That heaviness, or "force of gravity," that you feel when standing on Earth disappears when you hop into a different reference frame, in particular, any free-float frame. According to Einstein, that force of gravity can't be real because it doesn't appear in all reference frames. If a change of reference frame makes it vanish, then it can't be one of those deep, underlying, objectively real aspects of the world. Farewell to gravity as we know it!

This is absurd, you say. Of course there's a force of gravity! But to say that is to contradict the now generalized Principle of Relativity. To say that the force of gravity as you now know it is real is to say that one frame of reference—a frame at rest with respect to Earth—is a better or more appropriate place for making judgments about physical reality than is a free-float frame. But the essence of relativity is to deny that any reference frame has such special status. If something is here in one frame but not in another, then that something can't be real. Period.

What Is Gravity?

What you think of as gravity—the heaviness you feel standing on Earth or the force that accelerates a plummeting stone—is not real because it disappears in some reference frames (i.e., in any free-float frame). That doesn't mean there's no such thing as gravity; rather, the real gravity must be something that can't be transformed away with a change of reference frame.

I've argued that gravity disappears completely in a free-float reference frame, like an orbiting spacecraft or the little room of Figure 14.2b. Actually, that conclusion isn't quite true. Let's make the room very much bigger, as suggested in Figure 14.3. You, too, have become gigantic, with an arm span of several thousand miles. You perform the ball-drop experiment again, this time dropping not one ball but two, one from each hand. What happens? Each ball accelerates downward, falling on a straight-line path toward Earth's center. Of course, you and the giant room are also accelerating downward, so again the balls appear, at first, to float right next to your hands. Then something strange happens. The balls begin to approach each other, the distance between them shrinking as time goes by. You can see why this occurs: The falling room is now so

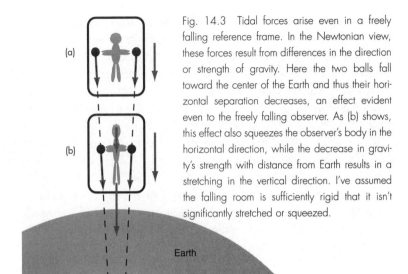

Fig. 14.3 Tidal forces arise even in a freely falling reference frame. In the Newtonian view, these forces result from differences in the direction or strength of gravity. Here the two balls fall toward the center of the Earth and thus their horizontal separation decreases, an effect evident even to the freely falling observer. As (b) shows, this effect also squeezes the observer's body in the horizontal direction, while the decrease in gravity's strength with distance from Earth results in a stretching in the vertical direction. I've assumed the falling room is sufficiently rigid that it isn't significantly stretched or squeezed.

large that the direction "down" is significantly different for the different balls. Each falls on a straight-line path toward Earth's center, but those paths converge, as Figure 14.3 shows. As a result, the balls drift closer together in the free-float reference frame of the huge, falling room.

Here's an effect that doesn't go away when you jump into a free-float reference frame. It's an effect you can use to distinguish between the situations of Figure 14.2. But it's a distinctly *nonlocal* effect—an effect you can't notice unless your experience spans a large region of space. (Actually, I should say, a large region of *space-time*, because you won't notice the effect even in the large room of Figure 14.3 unless you let the experiment proceed for some time.)

Newton would say that the drifting together of the two balls in Figure 14.3 is a subtle effect resulting from the variation of Newtonian gravity from place to place, in this case a difference in the direction of the Newtonian gravitational force at the locations of the two balls. In a related experiment, you could imagine dropping two balls, one at your head and one near your feet. Because Newtonian gravity is stronger closer to Earth, the lower ball falls with a slightly greater acceleration than the upper one; as a result, the distance between the balls would *increase* as they fell. A solid but deformable object, like a blob of Jello or even your body, would experience these effects as a squeezing sideways and a stretching vertically (Figure 14.3b). These subtle effects, known since Newton's time, explain the squeezing and stretching of Earth's oceans that we call tides. For that reason, they're called *tidal forces*. In Newton's view the tides aren't caused by gravity itself but by variations in gravity from place to place.

To Einstein, what Newton calls gravity doesn't exist, because it disappears in a free-float reference frame. Tidal effects, however, don't disappear, so they can claim to be real. For Einstein, those effects are the true manifestation of gravity. Newton says, "Tidal effects result from variations in gravity." Einstein says, "Tidal effects result from gravity itself." To Einstein, gravity in the Newtonian sense doesn't exist, so there's nothing to vary.

Curved Spacetime

So what *is* gravity? To Newton, it's a mysterious force that somehow reaches out across an immutable, unchanging, universal, three-dimensional space to influence distant objects. To Einstein it's much simpler. Gravity, Einstein says, is synonymous with the geometry of spacetime.

Imagine yourself a tiny being confined to the surface of a large, smooth sphere. Because you're so tiny, your day-to-day existence occupies only a small portion of the sphere. You study your world by doing the sorts of things you did in tenth-grade geometry. You draw triangles, for example, and discover that their angles add up to 180 degrees. You draw parallel lines, follow them for some distance, and find that they don't intersect (Figure 14.4a). You can take a journey halfway around the sphere and repeat these geometric experiments in another small region. You'll come to the same conclusion—that your world obeys the laws of tenth-grade geometry. This geometry is also called *Euclidean geometry*, after the mathematician Euclid, and so you declare that your space is Euclidean.

At least, it's *locally* Euclidean. The region in Figure 14.4a is so small that you don't notice the curvature of your world. Your meas-

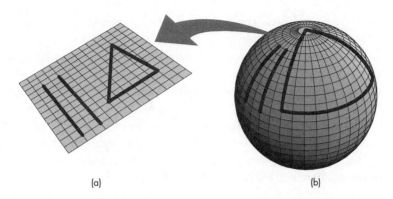

Fig. 14.4 (a) A small portion of a sphere is indistinguishable from a flat plane and obeys the laws of Euclidean geometry. Parallel lines never meet, and the angles of a triangle add to 180 degrees. (b) On a larger scale the sphere's surface is decidedly not Euclidean. Parallel lines approach and would eventually intersect. The large triangle shown has three right angles, which add to 270 degrees.

uring instruments just aren't sensitive enough to tell the difference between a truly flat world and a very large sphere. That's why Euclidean geometry works for you. (This situation is not so far-fetched. We all know that Earth is round, but that's because we're told so. Direct evidence for a round Earth is not at all obvious from everyday life, and if we're not airplane pilots or astronauts, most of us could get along just fine in our local communities by treating Earth as flat.)

If you explore a larger portion of your sphere world, then you begin to notice strange things. For example, you and a friend start out walking on perfectly parallel paths. You both continue walking absolutely straight, but after a while you notice that you're getting closer together (Figure 14.4b). Why? Because the straightest possible lines on the surface of a sphere, the shortest paths between any two points, are circular arcs centered on the sphere's center (in geography they're called *great circles*). Follow any two of those arcs and eventually they intersect (at two points, no less, if you go far enough). Or draw a really big triangle. I've shown one in Figure 14.4b that is so big it extends from equator to pole, and I've made it an equilateral triangle (all sides the same length). But its angles aren't the 60-degree angles of a Euclidean equilateral triangle; rather, they're each 90 degrees, and sum not to 180 degrees but to 270 degrees! Euclidean geometry just doesn't hold on this spherical world. The reason is obvious to those of us who can see the whole sphere from afar: it's because the sphere's surface is curved, not flat.

Confined to the sphere's surface, you might try to explain these strange results by proposing a mysterious "force" that tugs on you and your friend, pulling you away from "true" straight lines and causing your paths to converge. Similarly, the tape measures you use to stake out that big triangle might be deflected by the same force, accounting for the unusual angles. You might try to learn more about this force by varying your experiments. For example, replace your friend by an elephant and repeat the parallel-path experiment. Obviously, the same thing happens. You and the elephant gradually approach each other, despite the most meticulous effort at following straight paths. That gradual approach is just the same for the elephant as it was for your friend; therefore, you conclude that your proposed "force" has the same effect on all objects, independent of how massive they are. But your force idea is complicated and cumbersome, and invokes something that just can't be detected in a small, local region of your globe. How much simpler is the correct explanation! The geometry of this spherical world is not Euclidean; rather, it has curvature and that curvature accounts fully for strange effects like the gradual approach of objects on parallel, straight-line paths.

I've just presented an analogy for Einstein's conception of gravity. We live, says Einstein, in a four-dimensional spacetime. The geometry of spacetime exhibits curvature, and that spacetime curvature is gravity. Not "is a manifestation of gravity" or "is caused by gravity" or "causes gravity." No: spacetime curvature *is* gravity. Gravity is not some force that affects objects in spacetime. Gravity *is* no more and no less than the curved geometry of spacetime.

In a small, local region of spacetime—that is, in a local free-float reference frame—you don't notice gravity because spacetime curvature is negligible over small regions of space and time. Similarly, a small region on the surface of our hypothetical globe is essentially flat and obeys the laws of Euclidean geometry. In either case, one's world is simple when viewed *locally*. On a small patch of the sphere, you can use tenth-grade Euclidean geometry. Objects move in straight lines, parallel lines don't intersect, and triangles have 180 degrees. In a small, localized free-float reference frame in spacetime, physics is simple. The law of inertia holds, with free objects moving in straight lines at constant speed. It's only when you go beyond your local neighborhood that you notice deviations from simple geometry and simple physics.

A description of physical reality in a local free-float reference frame makes no mention of gravity. Objects at rest remain at rest, and objects in motion remain in straight-line motion at constant speed. Make that reference frame a space station orbiting Earth and this simple description remains true. There's no mention of Earth and its "gravity." Objects in the space station obey very simple physical laws because, locally, spacetime is flat. Here's a big philosophical advantage of Einstein's view: There's no such thing as "action at a distance." Rather, matter takes its "marching orders" from its immediate vicinity, that is, from the *local* geometry of spacetime. When that geometry is flat, as it always is in a small enough region of spacetime, those marching orders say to remain at rest if you're at rest, and to keep moving uniformly if that's what you're doing. What could be simpler?

Things get more complicated only when we consider larger regions of spacetime, large enough that spacetime curvature becomes noticeable. Then, as in Figure 14.3 (and its analog, Figure 14.4b), we notice the effects of that curvature on the paths of widely separated objects. We can call such effects "gravitational," but we could equally well call them "geometrical." Gravity is synonymous with the curved geometry of spacetime.

Locally, physics is always simple. The laws of physics work the same way in every small free-float frame of reference. But because of gravity—because of spacetime curvature—there is no one free-float frame that spans all of spacetime. The local free-float frame in one small region of spacetime is not the same as the local free-float frame in another region. For example, a freely falling elevator in New York is not the same free-float reference frame as a freely falling elevator in Bombay, on the other side of the Earth. That physics works equally well (and simply) in both elevators is the essence of special relativity.

What general relativity does is to provide the link between different free-float frames, giving a description of physics that is consistent across large regions of spacetime. As we'll soon see, general relativity also tells how and why spacetime is curved.

Before moving on to summarize general relativity, I want to counter an objection you might have to my sphere analogy. Aren't there "really" straight lines that cut through the interior of the sphere? Doesn't the big triangle "really" have 60-degree angles if I use those "truly straight" lines as its sides? Yes—but I want you to consider only the surface of the sphere. That's a two-dimensional surface because on it you can move in only two mutually perpendicular directions. Now, *you* can't imagine that sphere without seeing it curved in three-dimensional space. As the hypothetical tiny creature on the sphere, though, you can do experiments that demonstrate your world's curvature without ever leaving its surface. Walk in a straight line and eventually you come back to your starting point. Measure a big triangle and find that its angles add to more than 180 degrees. You just can't explain those happenings in a Euclidean world. You don't have to imagine three-dimensional space to recognize that the geometry of your world is not Euclid's geometry. In fact, you can give an accurate and consistent description of the sphere's curved two-dimensional surface without any need for a third dimension. The spherical surface's curvature is an *intrinsic* property of the surface itself, complete without any reference to a higher dimension. It's just that your three-dimensional brain finds it much easier to acknowledge that curvature if you picture the spherical surface as being curved in a third dimension.

Similarly, you and I have difficulty wrapping our minds around the idea that we live in a curved four-dimensional spacetime. Four dimensions are hard enough, and now they're curved as well? What are they curved in, a fifth dimension? No: just like the denizens of that sphere world, we can do experiments in our four-dimensional spacetime that tell us its geometry is not flat. That curved geometry is an intrinsic property of spacetime, and it doesn't require a fifth dimension. That curved geometry is gravity.

A Natural State of Motion

Roll a ball on that hypothetical sphere world or fly an airplane over the real Earth on the shortest path from San Francisco to Tokyo—a path that takes you by southern Alaska. In either case, your path isn't a Euclidean straight line because of your world's curvature, but in both cases it is the *straightest possible path* in that curved geometry. The ball rolling on the sphere world just naturally follows that straightest path, and so does the plane if the pilot doesn't take active steps to turn it. As long as there are no forces acting on them, both ball and plane follow the straightest possible path. Those straightest paths in curved geometry are called *geodesics.*

Einstein's law of motion for objects moving freely in the curved geometry of four-dimensional spacetime is similar. As long as no forces act on it, an object moves in the straightest possible path consistent with the geometry of spacetime in its immediate vicinity. In other words, objects follow geodesic paths in curved spacetime. That, in a nutshell, is half of general relativity—the half that tells how objects move. This law of geodesic motion ultimately covers everything from falling apples to planets and space shuttles and on to the overall behavior of the Universe as a whole.

Did I say "as long as no forces act"? Was I forgetting about the force of gravity? No: In Einstein's view, *gravity is not a force.* Of course not; it's just the geometry of spacetime. Objects move in the simplest, straightest possible paths when there are no forces acting. It's just that those paths aren't Euclidean straight lines, because spacetime is curved. In other words, because there is gravity.

In the broad historical context, we've just revisited the question I asked way back in Chapter 3: What is the "natural" state of motion? For Aristotle, the natural state of motion in the terrestrial realm was to be at rest as close as possible to Earth's center. Galileo and Newton swept away 2,000 years of Aristotelian thinking with their assertion that the natural state is, instead, uniform motion in a straight line. According to Galileo and Newton, uniform motion needs no explanation; only when motion *changes* should we look for a cause—that is, for a force. For Newton, one instance of changing motion is free fall under the influence of the gravitational force. Einstein built special relativity on the Newtonian premise, enlarging the Galilean relativity principle inherent in the Galilean/Newtonian view to include all of physics. Now, with general relativity, we broaden once more the idea of a natural state of motion. In general relativity, the natural state of motion is a geodesic, or straightest possible path, through spacetime. Such motion, which includes Earth's orbital motion around the Sun or the orbit of the International Space Station, needs no explanation in terms of "force" because it's the simplest, most natural motion possible. *Gravity isn't a force! It's the geometry of spacetime!* If an object isn't moving in a Euclidean straight line, that's to be expected in a spacetime whose geometry is not that of Euclid.

Note once again that gravity, or spacetime curvature, is a *global* effect. You'll never observe it locally, in a small enough free-float reference frame. Do experiments on the Space Station and the straight-line geometry of special relativity works just fine, even as spacetime curvature guides the station on a path whose spatial part circles the Earth. It's only when you look at the big picture, at how different local free-float frames are related, that you notice spacetime curvature, or gravity.

The Twins, Revisited

General relativity says that objects in free float follow geodesics in curved spacetime. We can use that fact to give a general relativistic description of the twins' star trip from Chapter 9. First I need to clear up one ambiguity. That straightest possible path in curved spacetime need not be the shortest. There are, in fact, two straightest possible paths between any two points on our sphere world or on the surface of the real Earth. Both are sections of the same circumference—a circle centered on the center of the sphere. For instance, you can get from San Francisco to Tokyo on that straightest path that takes you by Alaska and that's the shortest possible route. But if you go straight in exactly the opposite direction, you'll also get to Tokyo. That will be a longer way, around the whole rest of the circumference that you don't traverse on the Alaska route. Now, general relativity says that an object in free-float follows the straightest possible path through spacetime but that doesn't mean the shortest path. In fact, a kind of "principle of cosmic laziness" applies, in that the straightest path for a free-float observer in space-time is also the path on which the time between two events is the greatest. Greatest compared with what? Compared with other observers who also manage to move in such a way that they're present at the two events—observers whose motion is not free-float; that is, observers who experience forces.

Now let's apply this to the twins. We'll neglect Earth's weak gravity and consider that the earthbound twin is essentially in free float. Then we have two observers, the Earth twin and the space-traveling twin, who are both present at two specific events, the departure and the arrival of the ship on its round-trip journey. Whose clock records the longest time between these events? General relativity gives us the answer: the clock of the free-float twin, that is, the one who remains on Earth, in unaccelerated motion. The other twin's path through spacetime necessarily involves acceleration, and therefore forces, in order to turn around and make the journey a round-trip. So the space traveler's clock reads less time, and therefore she returns younger than her stay-at-home twin.

Figure 14.5 shows a Chapter-11-like spacetime diagram for the twins' situation. On the diagram are worldlines for three different observers: the stay-at-home twin, the traveling twin we discussed in Chapter 9, and a third observer who takes a different round-trip journey. As with any diagram in relativity, I should tell you the point of view (i.e., frame of reference) from which it's drawn. Here it's the Earth–star frame. In that frame the stay-at-home twin doesn't go anywhere in space, giving a worldline that's just a vertical segment along the time axis. The traveling twin moves away from Earth in space and forward in time; hence a diagonal worldline. Then she turns around at the star, marked by the sharp bend in her worldline, and returns to Earth. All the while she advances in time. Note that she returns to the same *place* but not the same *time* (of course not;

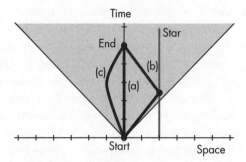

Fig. 14.5 Spacetime diagram for the star trip of Chapter 9, drawn from the Earth–star reference frame. Shown are worldlines for three observers who are all present on Earth at the ship's departure and again at its return: (a) the earthbound twin, (b) the twin who journeys to a 20-light-year distant star and returns, and (c) another observer who travels in the opposite direction. Tick marks are 10 years and 10 light-years apart on the time and space axes, respectively. Points mark the start and end of the journey and the traveling twin's turnaround at the star. Vertical gray line is the worldline of the star; Earth's worldline is the time axis itself.

the journey takes time). Remember that points on the spacetime diagram aren't places, but *events*. You might think the traveling twin's worldline is longer, and it looks longer on this diagram drawn on ordinary two-dimensional paper. But with the weird negative-sign Pythagorean theorem that applies in relativity, the time associated with her path is actually shorter than that of the stay-at-home twin. The third observer, who goes off to the left but returns to Earth just when the space-traveling twin does, records an intermediate time and therefore comes back older than the traveling twin but younger than the stay-at-home twin. By the way, you can tell from the worldline (how?) that this observer doesn't reverse abruptly and travels faster on return than on the outgoing trip. You can also see that the stay-at-home twin is the only one who can be present at both departure and return of the other observers' spaceships without having to undergo acceleration. For this twin, therefore, the time between the events is the longest possible among any observers present at both events.

What Curves Spacetime?

So spacetime is curved, and that curvature determines the natural, unforced motion of free objects. How does spacetime acquire its curvature? That's the other half of general relativity. The first half tells how matter moves in curved spacetime. The second half also involves matter, for it is, in fact, the presence of matter that curves spacetime. Actually, given the relativistic equivalence of mass and energy, it's the presence of matter or energy that curves spacetime.

Absent any matter, spacetime would be flat and objects would move in the truly straight lines of Euclidean geometry. But in the presence of matter or energy, the geometry of spacetime changes. Spacetime acquires curvature. The so-called field equations of general relativity tell quantitatively what curvature results from a given distribution of matter and energy. (Although I used the words "flat" and "Euclidean" here, even in the absence of matter spacetime still follows the weird, negative-sign Pythagorean theorem we've encountered before. So it isn't quite Euclidean, although its three spatial dimensions do obey Euclidean geometry.)

So spacetime and matter are engaged in a sort of cosmic tango. Matter (and its relativistic equivalent, energy) act on spacetime, giving it curvature. Curved spacetime, in turn, acts back on matter, telling it how to behave. Although the mathematical description of this relationship is complicated, its essence is simple. Where matter is densely concentrated, spacetime is more curved. When matter moves through spacetime, it does so in the straightest possible paths. That's it.

It's more difficult to picture this general relativistic relation between matter and spacetime than it is to imagine the forces and orbits of Newtonian gravitation. But it can be done, as in our sphere example, by giving up a few dimensions. Imagine a large sheet of rubber, stretched out horizontally. This two-dimensional surface is an analog for four-dimensional spacetime. Right now it's flat. Roll a little ball along it, and the ball follows a straight line.

Now place a heavy ball on the sheet. The sheet deforms, with the ball at the bottom of a depression caused by its own presence (Fig-

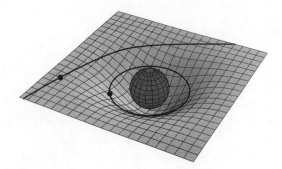

Fig. 14.6 A two-dimensional analogy for curved spacetime. The large sphere distorts a rubber sheet. Smaller objects moving on the sheet naturally follow curved paths. One object is moving fast enough that it gets deflected, then continues on; the other is trapped in a closed orbit around the large sphere. In both cases the motions result not from a force of attraction to the sphere but from the curved geometry of the rubber sheet.

ure 14.6). Now, you and I know the reason this happens is because old-fashioned Newtonian gravity pulls down on the ball. But that's not what this is about. The important point is not that something pulls the ball down but that the presence of the ball on the sheet, for whatever reason, distorts the sheet. The geometry of the sheet changes from flat to curved. That's just what happens to spacetime in the presence of matter.

Now suppose we roll a little ball along the curved sheet. Far from the large ball, the sheet remains nearly flat, and the little ball rolls in a straight line. As it approaches the large ball, the small one encounters substantial curvature and is deflected from its straight-line path. In fact, you can even put a smaller ball in "orbit" around the larger one, as shown in Figure 14.6.

Let's make the rubber sheet perfectly transparent and look straight down on it. All you see are the large ball and the small one. You study the motion of the small ball and conclude that it doesn't follow a straight line but gets deflected toward the larger ball. That deflection is minimal far from the large ball but becomes substantial closer in. You can even give the small ball just the right speed that it's deflected at a constant rate and circles endlessly around the large ball. You might well conclude that there is an attractive force between the two balls and that the force gets weaker with increasing distance between them. You would, of course, be reasoning like Newton, and discovering what you would happily call the "force of gravity."

But those of us who can see and touch the rubber sheet know better. There is no attractive force, no "force of gravity." There's only the rubber sheet, whose geometry is curved in the presence of the large ball. The amount of curvature depends on how close you are to the large ball. Small objects move across this curved sheet on simple paths, namely, the straightest lines possible in the curved geometry. This analogy, in a nutshell, describes general relativity. There is no "force of gravity." There's only spacetime, whose geometry is curved in the presence of matter. Objects move through this curved spacetime on the simplest, straightest paths possible in the curved geometry.

Note, by the way, that the actual curvature of the rubber sheet is not synonymous with the curved path of the small ball. You can see that it isn't by looking at how the sheet is distorted. Furthermore, my rubber-sheet example is incomplete because it shows only curvature in *space*, while gravity is the curvature of *spacetime*. So what

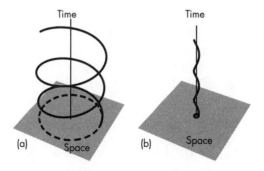

Fig. 14.7 Spacetime diagrams of Earth's orbital motion, showing two spatial dimensions and time. Helical paths represent Earth's worldline, its path in spacetime. Time and space scales are very different. In both versions the time axis spans 2.5 years, while in (a) the spatial extent is only about 20 light-minutes. Dashed circle shows Earth's orbit in space alone. In (b) the spatial scale spans about 250 light-minutes, and Earth's worldline shows more correctly as the nearly straight line our planet follows in the modestly curved spacetime of the Solar System.

does an object's path—say, Earth's orbit around the Sun—look like in curved spacetime? That's impossible to draw accurately, for the same reason you can't make an accurate map of the spherical Earth on a flat sheet of paper. So the picture I come up with cannot be a realistic depiction of Earth's motion in curved spacetime. In particular, the orbital path I show doesn't look at all straight or "straightest" any more than the "straightest lines" on the globe in Figure 14.4 look straight. Those caveats understood, take a look at Figure 14.7a. The orbital path—Earth's worldline, in the language of Chapter 11—is a spiral, because Earth goes around a circle in space while advancing straight into the future in time. Now, the diameter of Earth's orbit is about 200 million miles—that's 16 light-minutes, or about 30 millionths of a light-year. It takes Earth, of course, 1 year to complete 1 orbit. To show a few years' motion, then, I've had to use very different scales for the space and time measures in Figure 14.7a. If I had used the natural relativistic units of the spacetime diagrams in Chapter 11 (i.e., light-years for space and years for time, with each unit occupying the same physical distance on the diagram), then the time axis in Figure 14.7a would have to be about a thousand miles long or the spatial extent shrunk to some tens of millionths of an inch! I simply can't make a correctly scaled diagram without either going way off the page or shrinking the spatial scale so small that the circular orbit becomes microscopic. (I've tried in Figure 14.7b, but even there the scales are nowhere near correct.) Still, you get the point: Earth's path through spacetime is a very, very loose spiral; our planet advances much more in time (1 year) in each orbit than it does in space (an orbital circumference, or 200 millionths of a light-year). Drawn to scale, that spiral is very nearly a straight line—an indication that spacetime curvature in Earth's vicinity, although enough to produce our planet's circular orbit, is just not very great.

Curved spacetime, geodesics, matter and energy—in general relativity we have a radically different understanding of gravity, an understanding that arose almost single-mindedly in the brain of Albert Einstein. Philosophically, Einstein's theory of gravity is light-years different from Newton's. What does it tell us physically about our Universe? We'll go there in the next chapter.

INTO THE BLACK HOLE

● ● ●

General relativity certainly provides a philosophically different picture of gravity, but does it tell us anything new physically? Do planets orbit, or apples fall, differently in the general relativistic description of gravity? You might at first think not. After all, old-fashioned Newtonian gravity does such a good job predicting motion on Earth and throughout the Solar System that if general relativity's predictions were significantly different they would have to be wrong!

In fact, the predictions of general relativity do differ from those of Newton. On Earth, and indeed everywhere in our Solar System, those differences are so slight as to be almost immeasurable. Almost, but not quite. Very sensitive tests can distinguish Einstein's predictions from Newton's. Newtonian gravitation works fine for everyday tasks like launching a space shuttle, sending a spacecraft to Jupiter, or predicting eclipses. But when the utmost precision is needed, and when instruments are capable of measuring with great precision, then we can and sometimes must use general relativity to predict the effects of gravity.

Einstein and Newton differ only slightly in regions where gravity is weak. I'll give a precise definition of what I mean by "weak gravity" and "strong gravity" later in this chapter. Suffice it to say for now that nowhere in our Solar System—even in the vicinity of the Sun itself—is gravity at all strong. The same is true around most ordinary stars and planets that orbit them, but there are other places in the Universe where gravity is truly strong. There, Newtonian gravitation fails dramatically and the predictions of general relativity once again play havoc with our commonsense notions of time and space.

Errant Orbits

By the nineteenth century, Newtonian gravitation reigned supreme throughout the Solar System. Meticulous calculations of the gravitational effect not only of the Sun but also of the other planets had accounted for nearly every detail of the observed motions of the planets. But there was one tiny discrepancy in the motion of the planet Mercury, a discrepancy that remained even when the effects of the other known planets were included.

In an idealized Newtonian universe containing just the Sun and one planet, the planet's orbit would be a perfect ellipse that closes back on itself and repeats exactly forever (Figure 3.2 showed such an orbit). The real Solar System is more complicated, largely because of the gravitational effects of the planets on each other. We can subtract out those effects and ask if, in their absence, a planet's orbit would be the ideal, closed ellipse. For Mercury, the answer is "not quite." Mercury's ellipse doesn't quite close, meaning that the long axis of the ellipse rotates slowly with time, as shown in Figure 15.1. "Slowly" is the operative word here, for in 100 years the orbit swings through a mere 43 seconds of angle. You probably don't have a good feel for a "second of angle," but you know a "degree of angle" as a very small angle indeed; there are 360 degrees in a full circle, 90 in a right angle. One second of angle is a minuscule 1/3,600 of a degree. So the orientation of Mercury's orbit changes by only about 43/3,600 of a degree in a century, or just over a tenthousandth of a degree per year! That's a pretty small angle, and its measurement is all the more impressive because Mercury's orbit is not the obvious ellipse of Figure 15.1, but is nearly circular. Nevertheless, astronomical observations are accurate enough that nineteenth-century astronomers were confident in their measure of Mercury's anomalous orbital motion. They called the phenomenon "the precession of the perihelion of Mercury," where *precession* is a

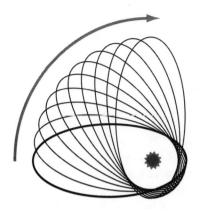

Fig. 15.1 Orbital precession. In Newtonian gravitation, a planet's elliptical orbit would repeat forever (thick ellipse), but general relativity predicts that the orientation of the orbit should change with time—an effect observed in Mercury's orbit. The precession shown here is highly exaggerated, and Mercury's orbit is also much closer to circular.

slow rotation like the change in the orientation of a spinning top and *perihelion* is the point on the orbit where the planet is closest to the Sun. You can see from Figure 15.1 that precession of the orbital axis carries with it that point of closest approach.

Mercury's perihelion precession posed a quandary for nineteenth-century physicists and astronomers. Because it was such a small effect and because Newtonian gravitation was so spectacularly successful, most believed it could be explained in the Newtonian context as the effect of an as yet undiscovered planet, an unseen moon of Mercury itself, or a swarm of asteroids. A few tried to modify Newton's law of gravity, but without much success or theoretical justification.

Then, in 1915, as he was putting the finishing touches on his general theory of relativity, Einstein tackled the question of Mercury's orbit. To his profound delight, he found that his theory predicted a perihelion precession of 43 seconds of angle per century! This result was a brilliant confirmation of what was otherwise an abstract, mathematical exercise in formulating Einstein's new description of gravity as the curvature of spacetime.

So one difference between Einstein's and Newton's theories is that, in Einstein's theory, planetary orbits don't quite close. This effect is hardly noticeable in the weak gravity of our Solar System, and it's no wonder it was discovered in Mercury—closest to the Sun and thus the planet subject to the strongest (but still very weak) gravity. Philosophically, the explanation of Mercury's perihelion precession in terms of general relativity represents a profound leap in our understanding of gravity. Physically, though, it's hardly a big deal, again, because gravity in our Solar System is so weak.

An Astrophysical Interlude

Observed differences between Newtonian gravitation and general relativity remained mostly small, marginal effects through the first two-thirds of the twentieth century. We'll soon look at some of the other classical tests that affirmed Einstein's theory using very sensitive measurements on Earth or elsewhere in our Solar System. But in the 1960s things began to change. Astrophysicists discovered amazing objects whose gravity was really strong in the sense I'll define shortly. The first such objects discovered were stars that, having expended the nuclear fuel that kept them shining, had collapsed under the influence of their own gravity to form one of two remarkable stellar endpoints. *Neutron stars*, the somewhat less bizarre of these two "dead star" possibilities, pack roughly the mass of the Sun into a region only a few miles across. Under the resulting intense gravity, electrons and protons collapse together to make neutrons, hence "neutron star." The neutron star is somewhat like a giant atomic nucleus, so dense that a teaspoon of neutron-star matter would weigh more than 100 million tons. An unusual kind of pressure arising from all those neutrons keeps the star from collapsing further, but there's an upper limit to the mass neutrons can support. Above this limit, the burned-out star must collapse completely, forming a *black hole*. Much more on black holes later—they're in this chapter's title—but for now suffice it to say that massive stars collapse at the ends of their lives to form either neutron stars or black holes. In both cases, gravity in the immediate vicinity of the collapsed star is most definitely strong. (By the way, our Sun is not sufficiently massive for such a dramatic fate; it will eventually become a white dwarf star, with its mass packed into roughly the size of the Earth.)

It was in the 1930s that physicists first used general relativity to predict the theoretical possibility of neutron stars and black holes, but it wasn't until the 1960s that these remarkable objects were actually discovered. I won't go into the details of how we detect neutron stars and black holes, except to note that they often occur in binary-star systems and are observed through their effect on a companion star. (You may recall from Chapter 5, where observations of binary stars showed that the speed of light does not depend on the speed of its source, that about half the stars in our galaxy are in binary systems.) In 1974 astrophysicists Joseph Taylor and Russell Hulse made a remarkable discovery: a binary system containing not one but two neutron stars in very close orbits. Each has about 50 percent more mass than our Sun and their orbits are decidedly elliptical. At closest approach, they're only about half the Sun's diameter apart. With such tight orbits, each completes a full orbit of the other in a mere 8 days. The gravity each neutron star experiences is much stronger than anything in our Solar System, and the effects of general relativity are much more pronounced. Thus this unusual binary system provides a laboratory for studying general relativity.

How do we know so much about the Taylor–Hulse binary system? Because neutron stars, in addition to being dense, often rotate at high speeds. This rapid rotation occurs for the same reason that a figure skater or ballet dancer goes into a rapid spin by pulling in his or her arms. As a large star collapses to a neutron star, its initially slow spin gets amplified as the matter of the star "pulls in" to occupy a much smaller space. (The physics term for this is "conservation of angular momentum.") Furthermore, neutron stars have strong magnetic fields, and these fields channel outgoing electromagnetic radiation—light, radio waves, x-rays—into narrow beams that sweep through the cosmos as the star rotates. When a beam sweeps by Earth, our telescopes or x-ray detectors record a pulse of radiation. As the neutron star spins, we get one such pulse for each rotation; for that reason we call these spinning neutron stars *pulsars*. By timing the pulses, we can measure the spin rate with amazing accuracy; in the Taylor–Hulse system, for example, the spin rate of one neutron star was measured to be 16.940539184253 rotations per second. (This measurement was made in 1986, and the rate has changed since then—more on this later!)

Now, as I noted in Chapter 5, the motion of stars in a binary system toward or away from us changes the frequency or color (but not the speed!) of the light we receive. Similarly, the pulse rate we measure depends on whether a pulsar is moving toward or away from us. (This phenomenon, called the *Doppler effect*, also occurs for sound and is probably familiar to you. For example, when you stand near a highway, you hear a high-pitched sound as a truck approaches, then a low-pitched sound as the truck passes and heads away from you: "aaaaaaaaaeeeiiiooooooooo." The high pitch occurs because wave crests pass you more often as the truck approaches; the low pitch because they pass less often as the truck recedes. The same idea holds for light-wave crests or pulsar pulses.) So by timing a pulsar's pulses, we can keep track of its orbital motion. Even though the system is way too distant for us to see the individual neutron stars, pulse timing allows us to construct a picture of the orbital motion. Because the pulse rate is known with such precision, that picture is very accurate.

What's all this got to do with general relativity? With objects as massive and as close as those in the binary pulsar, general relativity predicts a precession of the orbital axis by more than 4 degrees per year, some 35,000 times Mercury's paltry 43 seconds of angle per century. Given our accurate picture of the pulsar's orbit, derived from timing its pulses, the orbital precession is obvious and easy to measure. The result, from more than 20 years of observations, is a precession that's right in line with the general relativistic prediction.

So here's one way in which general relativity and Newtonian gravitation differ: elliptical orbits don't quite close, but precess at a rate that depends on how strong gravity is at the location of the orbit. In our Solar System, gravity is so weak that this precession is barely noticeable. Elsewhere in the cosmos, though, nature provides remarkable systems—here the binary pulsar—where general relativity's deviation from Newtonian physics is much more obvious. That will be the story of the rest of this chapter: we identify a general relativistic effect and confirm it early on with a very sensitive terrestrial or Solar System–based experiment. Later, toward the end of the twentieth century, astrophysics provides much more dramatic confirmation of the same effect. Today, in the twenty-first century, general relativity is a solidly verified theory, a working tool of many astrophysicists.

Light Bends!

In the previous chapter I introduced the Principle of Equivalence as the heart of general relativity. This principle says that no experiment can distinguish the state of free float—motion under the influence of gravity alone—from truly uniform motion in the complete absence of gravity. Figure 14.2 made that point very clear for a simple mechanical experiment like dropping a ball. Einstein extended the equivalence principle to all of physics, including electromagnetism and the behavior of light. He then turned his reasoning around and used the principle to deduce a new physical phenomenon, namely, the bending of light in the presence of gravity.

Consider once again the little room of Figure 14.2a, way off in intergalactic space where there's no gravity whatsoever. Suppose we mount a flashlight on one wall and shine the light directly across the room. The light goes straight across the room to make a bright spot on the opposite wall, directly opposite the flashlight (Figure 15.2a).

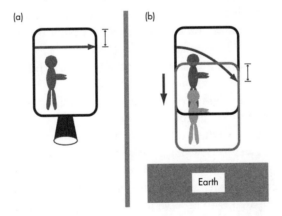

Fig. 15.2 (a) Light shined across the little room in gravity-free space hits the opposite wall. (b) The freely falling room is equivalent, so here the light should hit the opposite wall the same distance below the ceiling as it did in (a). But the room is accelerating downward, so the light's path must bend. The room is shown twice, first when the light departs the left wall and again (lighter image) when it hits the right wall.

Now switch to the situation of Figure 14.2b, where the room was shown falling freely toward Earth. Einstein, with his Principle of Equivalence, says that this situation is indistinguishable from that of Figures 14.2a and 15.2a. That means any experiment we do in the falling room must have exactly the same outcome that it did way off in intergalactic space. If it didn't, we would have a way to distinguish the two situations. So if we turn on the flashlight here, the light must still hit the wall directly opposite. In either situation, an observer in the room sees the light go through the room in a straight path to the opposite wall. Now think about what this looks like to someone on Earth watching the falling room. To the earthbound observer, light leaves the flashlight at the left side of the room. By the time it gets to the opposite wall, the room has fallen some distance, yet the light still hits the wall right opposite its source. How can that be? Only if the light, too, has "fallen." That is, the light itself must describe a curved path, as suggested in Figure 15.2b.

Einstein's conclusion that gravity must affect light shows the power of the equivalence principle in deducing new physical phenomena. Later, when he had formulated the full general theory, Einstein had another explanation: light, like matter, moves in the straightest possible paths through spacetime. But because spacetime is curved near massive objects, those paths are not the true straight lines of tenth-grade geometry. Incidentally, Einstein's final form of general relativity predicts a bending of light twice as great as his earlier equivalence-principle argument would imply.

So the path of a light ray should bend as the light passes a massive body. To test this idea, astronomers sought to measure the bending of light passing close to the most massive body in our Solar System—the Sun. The idea was to view one or more stars at a time when their light had to pass close to the Sun on its way to Earth (Figure 15.3). Astronomers could then measure changes in the stars' apparent positions as compared with observations made when the Sun was not in the same general direction as the stars. From those changes they could calculate the angle by which the Sun's gravity had deflected the light.

There's a problem with this approach: To view stars that appear in the sky near the Sun, one has to observe the stars in the daytime, but stars aren't visible in the bright daytime sky. So the only way to

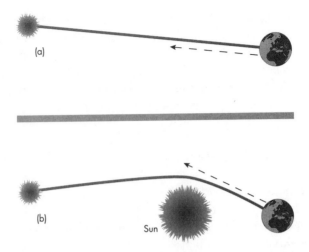

Fig. 15.3 (a) Light from a distant star comes directly to Earth. (b) When the Sun and star are in the same region of the sky, the Sun's gravity deflects the starlight. Observers on Earth then see the star at a different apparent position in the sky. Dashed lines mark the directions observers would look to see the star. The deflection shown is greatly exaggerated.

make this observation is during a total eclipse of the Sun. Ideally, one would like an eclipse that occurred at a time when many bright stars would appear near the Sun. Now, Einstein made his final calculation of the bending of light in 1916. Fortunately, an eclipse was to occur on May 29, 1919, and on that date it happens that many bright stars appear near the Sun. So the 1919 eclipse would provide an excellent test of Einstein's prediction.

Sir Arthur Stanley Eddington, a prominent British astronomer and early student of Einstein's work, together with colleagues began planning an expedition to the 1919 eclipse. Unfortunately World War I intervened. Eddington, a devout Quaker, sought exemption as a conscientious objector to military service. After several appeals, his exemption was finally granted, in part on the grounds that he was essential to the upcoming eclipse expedition. After the war's end, the British mounted two expeditions to view the 1919 eclipse, one to Brazil and the other, with Eddington its leader, to Principe Island off the African coast. After careful analysis of the observations, Eddington reported the mathematical details to a joint meeting of the Royal

Society and the Royal Astronomical Society: both eclipse sites confirmed Einstein's prediction. It was this result, not the special theory of relativity or the Mercury perihelion precession, that catapulted Einstein to fame. The *London Times* trumpeted the news with the headline "Revolution in Science . . . Newtonian Ideas Overthrown." Einstein spent the rest of his life in the public spotlight.

Gravitational Lenses

Like the precession of Mercury's perihelion, the bending of light remained through most of the twentieth century an obscure and very subtle effect significant only in helping confirm the general theory of relativity. Einstein himself saw greater possibilities, suggesting that massive bodies deep in space might provide much more dramatic bending of light—enough to produce distorted and even multiple images of more distant objects. Figure 15.4 shows how this works. Here, a massive galaxy lies between Earth and a quasar. *Quasars* are distant objects so bright they can outshine an entire galaxy, yet small enough that they appear as pointlike sources of light (more on quasars later). Light from the quasar bends so much

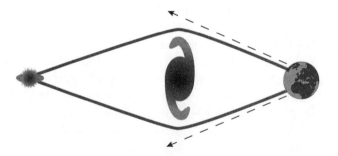

Fig. 15.4 Gravitational lensing. A massive galaxy lies between Earth and a distant quasar. Light from the quasar takes several paths around the galaxy, resulting in multiple images of the quasar as viewed from Earth. Dashed lines show two different directions in which the quasar is visible, resulting in two distinct quasar images. Additional images, or a continuous ring, would appear in three dimensions.

in passing the galaxy that an observer on Earth can see the quasar by looking in either of two directions—above or below the galaxy in Figure 15.4. The result is two images of the same object! In a telescopic photograph, these appear as two distinct objects at different positions on the photo. Actually, things get more complicated; in three dimensions the situation of Figure 15.4 would result in the quasar's image being smeared into a circular ring, provided the quasar were directly behind the galaxy and the galaxy perfectly circular. Absent this perfectly symmetric situation, the usual result would be several images of the same quasar.

It was not until the 1970s that astronomers firmly identified such *gravitational lenses.* By now, numerous examples of galaxies lensing quasars have been discovered. One of the best known is the "Einstein cross," a constellation of four images of the same quasar, shown in Figure 15.5a. In other examples, distant galaxies appear

(a)   (b)

Fig. 15.5 Hubble Space Telescope images showing gravitational lensing. (a) The Einstein Cross includes four images of the same quasar, gravitationally lensed by a massive galaxy. The central bright spot is the nucleus of the galaxy, with the four quasar images around it. Image quality is limited by the resolution of the Hubble Telescope. (b) Here an entire cluster of galaxies acts as a gravitational lens for more distant galaxies, producing distorted, arclike images. Some of these are multiple images of the same distant galaxy. The lensing galaxies are visible as the larger, brighter objects. (Credits: (a) NASA, J. Westphal, W. Keel; (b) NASA, A. Fruchter, and the ERO team [STScl, ST-ECF].)

smeared out and distorted as their light passes a nearer galaxy (Figure 15.5b). Study of these images gives information about the lensing object as well as the distant galaxies.

Today, gravitational lensing is more than a novel phenomenon that confirms Einstein's remarkable insight. It's also become a tool for exploring the cosmos. In Chapter 1, I described how a massive galaxy cluster some 2 billion light-years from Earth acts as a giant telescope, concentrating the light from more distant objects that would otherwise be too dim for us to detect with human-made telescopes. Lensing also gives astrophysicists insight into the overall structure and evolution of the Universe by providing distances to remote, multi-imaged quasars and estimates of the cosmic acceleration (more on this in the final chapter). Closer to home, gravitational lensing occurs in our own galaxy when a massive object passes in front of a star. As seen from Earth, the star appears to brighten briefly due to the focusing of its light by the object's gravity. Such *microlensing* has become an important tool for astronomers searching for the "missing mass" known to comprise much of the matter in the Universe. The technique might also help astronomers discover extrasolar planets by the microlensing effect as an unseen planet passes in front of its star.

In describing gravitational lenses, I've been using language like "light bends" and "focusing of light by gravity." That language helps explain the phenomenon, but it isn't quite relativistically correct. What's actually happening, of course, is that light is going in the straightest possible lines in curved spacetime. It's just that those aren't the ideal straight lines of tenth-grade geometry. Sometimes— as when multiple images form—the geometry of spacetime is such that there's more than one "straightest line" between a distant object and Earth.

Warping Time

In general relativity, gravity is the geometry of space*time*. So it should come as no surprise that time itself is affected by the presence of massive objects that warp spacetime. Once again it was the equivalence principle that led Einstein to this new phenomenon. He imagined the situation of Figure 15.2, but this time with the flashlight shining from floor to ceiling. In the frame of Figure 15.2a, with gravity absent, the light reaches the ceiling unchanged; in particular, it's the same color as it was at its source. According to the equivalence principle, that must also be the case in the equivalent freely falling frame of Figure 15.2b. To an observer on Earth, though, the room is accelerating downward as the light makes its journey from floor to ceiling. So when the light reaches the ceiling, the room is moving faster than it was when the light was emitted. That means an observer at the ceiling should see a Doppler shift, toward higher frequency and bluer color. But the equivalence principle says that doesn't happen, so there must be a redshift, associated with the presence of gravity, that cancels the blueshift. Observers who aren't accelerating downward would still see this redshift even though there would be no blueshift. So if I'm above Earth or some other gravitating body, and I observe light emitted at that body, then when I receive the light it's redshifted—that is, it has a lower frequency—compared to what an observer right at the light source would see. *Gravitational redshift* is one name for this phenomenon, but a deeper name is *gravitational time dilation*.

Why "time dilation"? Because the vibrations of a light wave, like any other periodically repeating phenomenon, provide a measure of time. In fact, our most accurate clocks—the atomic clocks that establish world time standards—use as their ticks the frequency of light waves emitted by particular atoms. So in the frequency and wavelength of light waves we have a measure of time itself. If I'm looking down at a source of light, the light reaching me has a lower frequency than it would if I were right beside the source. That is, the time interval between crests of the light wave is for me stretched out. But light frequency is simply a measure of the underlying passage of time; therefore, time intervals between *any* events occurring down at the light source also appear to me stretched out. That includes the interval between ticks of an ordinary clock; in other words, if I look down on a clock, I see it running slow compared to a clock right next to me. That's gravitational time dilation. As with the time-dilation phenomenon we encountered in special relativity, gravitational time-dilation isn't about light or clocks—it's about time itself. Time runs at different rates depending on where you are in relation to a massive gravitating object.

Is the lower clock really running slow? If I go down there and stand next to it, its timekeeping will seem perfectly normal. Of course; the clock is in a perfectly legitimate frame of reference and so it works normally. But compared with the clocks of an observer located higher up, the lower clock really does run slow. This time-dilation effect, unlike the time dilation of special relativity, is not reciprocal. That is, a clock lower down really does keep time more slowly compared with one higher up, and the clock higher up really is running fast compared with the lower clock.

Does gravitational time dilation actually happen? It does and it's been measured. Here on Earth the effect, like all manifestations of general relativity, is subtle and difficult to detect. However, in a famous 1960 experiment, physicists managed to measure the difference in timekeeping rates of clocks at the top and bottom of a 74-foot tower at Harvard University. Their clocks were atomic nuclei emitting radiation whose changes in wavelength could be detected with exquisite precision. The result—a shift of only about a thousandth of a trillionth of the original wavelength—verified general relativistic time dilation for the weak gravity at Earth's surface. In the 1971 experiment I introduced in Chapter 1, scientists flew atomic clocks around the world and compared their timekeeping with an atomic clock left behind. The combination of special relativistic time dilation associated with relative motion and reduced gravitational time dilation from higher altitude were fully consistent with relativity. (See "Around the World Atomic Clocks," in the Further Readings, for details of this experiment.) Today, the Global Positioning System (GPS) times signals from a constellation of orbiting satellites to provide precise locations anywhere on Earth. So accurate is GPS that if the satellites' atomic clock times weren't corrected for gravitational time dilation, the system would soon be off by a matter of miles!

As always with general relativistic effects, it's in the astrophysical realm that gravitational time dilation is most obvious. Even the Sun's weak gravity produces a measurable effect. Decades before the earthbound Harvard experiment, astronomers had measured the gravitational redshift of light from a white dwarf star, which boasts the Sun's mass crammed into the size of the Earth. At the surface of such a dense object, gravitational time dilation is some 30 times greater than that of the Sun. Finally, today's astrophysicists routinely observe substantial time dilation in the strong gravity around neutron stars and their even more bizarre cousins, the black holes.

Black Holes

What goes up must come down, right? No! Throw a ball straight up as hard as you can and it eventually slows, stops, and returns. If you could throw it fast enough—for an object thrown from Earth's surface, "fast enough" is about 7 miles per second—the ball would have enough energy to escape Earth's gravity altogether and would travel outward forever without stopping. That's because gravity weakens so rapidly with increasing distance that escape to an infinitely great distance does not require infinite energy. The 7 miles per second you'd need is called the *escape speed* for Earth's surface. Although the human arm can't propel anything at escape speed, rockets can. Spacecraft traveling to the outer planets, for example, leave Earth's vicinity at greater than escape speed. Pioneer and Voyager spacecraft even exceed escape speed for the Sun, meaning they'll eventually leave our Solar System and spend the foreseeable future drifting through the galaxy.

What determines escape speed from a given location? Ultimately, it's the strength of gravity at that point. For the surface of a planet or star, that's set by the mass and size of the object. Cram more mass into a given-size object and escape speed goes up. Shrink an object of fixed mass and again escape speed goes up. So imagine compressing Earth to ever-smaller sizes. Escape speed from the surface of the shrinking planet rises from its current 7 miles a second to ever-higher values. It gets harder and harder to "throw" a spacecraft or other object forever outward, but with sufficient energy and advanced technology, it remains possible. Possible, that is, until escape speed reaches the ultimate value, namely, the speed of light. For Earth, that would happen when the entire planet was a little under an inch in diameter. That's right—you, me, Mount Everest, New York City, all the water in the oceans, all the continents, the liquid and solid cores of the planet—all crammed into a space smaller than a Ping-Pong ball. If this seemingly impossible compression occurred, then we would have an object so dense that not even light could escape. That's a black hole.

Our incredible shrinking Earth, compacting until its escape speed approaches the speed of light, finally provides a solid definition of the terms "strong" and "weak" gravity that I've been using throughout this chapter. Strong gravity exists where escape speed is an appreciable fraction of the speed of light, c. Weak gravity means escape speed is far less than c. Earth's and Sun's escape speeds, at 7 and 380 miles per second, respectively, are far less than the 186,000-mile-per-second speed of light. Gravity everywhere in our Solar System is weak. At the surface of a typical neutron star, however, escape speed is about two-thirds that of light. This is strong gravity! Shrink that neutron star even a little bit and it will collapse to a black hole with escape speed c—the ultimate in strong gravity.

A black hole is a remarkable object. Our current understanding of physics suggests that once an object has been squeezed to black-hole size, there's no force in the Universe that can prevent its further collapse to a single point of infinite density. This infinite conclusion may change somewhat when we finally learn how to merge general relativity with quantum physics, the theory that describes matter at the atomic and subatomic scales. Even so, black holes will remain objects in which matter is compressed to a near point of incredible density. Surrounding this point is a spherical surface called the *event horizon*, which bounds the region within which the escape speed exceeds the speed of light. No light can escape from within this region, making it a true horizon. Those of us on the outside can never see in, past the horizon. There's simply no way for us to get information about events occurring within the horizon, hence the name *event* horizon.

Because light can't escape a black hole, and since no material object can go faster than light, that means nothing whatsoever can escape the hole. That fact makes black holes remarkably simple objects. From the outside, black holes exhibit very few distinguishing properties. Most significant is their mass—the total amount of matter and energy that has fallen across the horizon. It doesn't matter whether that mass-energy was in the form of stars, planets, people, interstellar dust, mice, water, light, or whatever. Once it's across the horizon, we can't know anything about it and so all that matters is the total mass. That mass determines the size of the event horizon and the gravitational influence the hole has on the surrounding Universe. If the infalling matter has electric charge, the black hole, too, will be charged, and the charge will be felt outside the horizon. If the infalling matter has rotational motion, the black hole will itself be spinning in a way that influences spacetime outside the horizon. But that's it: mass, spin, and electric charge are the only properties that distinguish black holes.

People often picture a black hole as sucking up all the matter in its vicinity. That's a misconception, because a black hole's gravitational influence is the same as that of any other object with the same mass. Far from the hole, matter will orbit in essentially Newtonian elliptical orbits determined by the hole's mass alone. If Earth suddenly collapsed to a black hole, for example, the Moon would be completely unaffected and would continue in its orbit about the Earth-mass black hole. It would not suddenly be sucked in any more than the Moon or a satellite is sucked to Earth by the planet's gravity. The only objects that strike Earth are those that are on a collision course with our planet or are close enough that Earth's gravity deflects them toward a collision. The same is true with a black hole; only matter that comes very close to the event horizon actually falls through it and because typical event horizons are very small, such a course is quite improbable. That means an isolated black hole will swallow matter at a rather low rate. On the other hand, a hole surrounded by a dense aggregation of matter—as in a binary star system or near a galactic center—will generate a substantial inflow of matter. More on this when we consider real black holes out there in the cosmos.

The notion of a black hole behaving like a cosmic vacuum cleaner does, however, have some merit. That's because the event horizon is a one-way street; matter that crosses the horizon can never re-emerge. So a black hole only grows in mass. Again, it doesn't do so by inexorably pulling in everything around it; rather, whatever happens to fall into the hole simply doesn't get out.

Actually, even that conclusion has to be tempered. In a remarkable quantum-physics process first envisioned by Stephen Hawking, black holes can actually lose mass by evaporation involving particles created in the vacuum just outside the event horizon. For astronomical-size black holes, this process is so feebly slow as to be completely negligible, but it might play a role in the very long-term evolution of the Universe.

Journey into a Black Hole

Black holes provide the ultimate time warp. That's because gravitational time dilation becomes infinite at the event horizon of a black hole. To see what this means, imagine that you and I are positioned a fixed distance far from a black hole. Each of us has a clock, and we've carefully synchronized them to read the same time. You then proceed to fall toward the black hole while I remain behind. As you fall, gravitational time dilation slows your clock relative to mine. So I see your clock, and indeed all manifestations of time, running slower and slower as you approach the event horizon. The movements of your body appear to slow. If I monitor your heartbeat, it slows. You appear to age more slowly. Time itself is running slower for you, in the warped spacetime around the hole, than it is for me. As you approach the event horizon, I see your time running ever slower until, at the horizon itself, it stops completely! I never see you quite reach the horizon because by the time that happens, it's the infinite future for me! So if I'm watching matter falling into a black hole, it appears to me to "freeze" just outside the event horizon. In my time frame, infalling matter never crosses the horizon. For that reason Russian astrophysicists coined the term "frozen star" to describe black holes forming from collapsed stars.

By the way, if you decide to abort your black-hole journey and blast away using a powerful rocket, you can of course do so as long as you haven't yet crossed the horizon. Because your time has been running so slowly near the hole, you'll return to find me much older than you or even long gone as you come back to your starting place centuries, millennia, or even farther in the future. But for you only a matter of hours or days might have elapsed. The numerical details depend on the mass of the hole and how close you let yourself get. In any event, here's another way—like the twins example of special relativity—for you to leapfrog your way into the future. Again, there's no going back if you don't like what you find!

So far I've described your black-hole journey largely from my viewpoint as an observer distant from the hole. But what's it like for you? Assuming you don't decide to return, you'll find yourself descending toward the hole and right through the event horizon in a finite and possibly even very short time. Will you experience your clock slowing, your movements becoming languid, and your body remaining forever youthful? Not at all! That would violate the Principle of Relativity. Your freely falling or free-float reference frame is a perfectly good one for doing physics, and your clock, your body, and all other physical processes should seem perfectly normal to you. Will something dramatic happen at the moment you cross the event horizon? No! Again, the relativity principle asserts that the laws of physics work just perfectly in your free-float reference frame, so if something odd happened right at the horizon, that would violate relativity. You wouldn't even know when you had fallen past the point of no return.

There is, however, a practical constraint on your imaginary black-hole journey. For a black hole formed from a collapsed star (more on this in the next section), the curvature of spacetime would become so great that your body would be torn to pieces before you crossed the horizon. Figure 14.3 showed how the spacetime curvature that constitutes Einstein's gravity results from what Newton would describe as differences in Newtonian gravity from place to place. As Figure 14.3 suggests, the consequence in either Einstein's or Newton's picture is a force that stretches a freely falling body in the vertical direction and compresses it horizontally. For a star-mass black hole, those forces become fatally large well outside the event horizon. For black holes of a billion star masses, though—holes now believed to inhabit the centers of some galaxies—the tidal forces on a human body would be negligible at the event horizon. You would fall freely across the horizon of such a hole with no sense whatsoever of anything unusual.

Once inside *any* black hole, however, you'd be drawn inexorably toward the singular point at the hole's center. No force in the Universe is strong enough to resist the pull or, in general relativistic terms, to prevent your worldline through spacetime from intersecting the singular point. Sooner or later (sooner for a lower-mass hole, later for a larger one) spacetime curvature will become significant over the size of your body and you'll be destroyed.

Do Black Holes Exist?

Black holes remained figments of theorists' imaginations until the late twentieth century. Then, with the advent of space-based astronomy in the 1960s, astrophysicists suddenly had new windows on the Universe. Particularly significant in the discovery of black holes was the x-ray window. X-rays, similar to those used in medical imaging, don't penetrate Earth's atmosphere. When astrophysicists first turned satellite-borne x-ray detectors on the cosmos, they were surprised to find a number of bright, pointlike x-ray sources. Many of these proved to be binary stars in which x-rays were produced in the intense heat generated as gas flowed from a large, visible star to an unseen companion. Those unseen companions became the first candidates for black holes.

Why should a black hole become a cosmic x-ray source? Because black holes, and their neutron-star cousins, are so deep gravitationally that infalling matter accelerates to enormous speeds. If that matter is gaseous, friction in the flowing gas generates very high temperatures—so high that instead of glowing red-hot like a stove burner, or white-hot like our Sun, the infalling gas "glows" in x-ray "light." (Recall from Chapter 4 that x-rays are just another form of electromagnetic waves, distinguished from visible light by their much shorter wavelength.) Some x-ray sources showed rapid periodic fluctuations, indicating that the unseen companion was a rapidly spinning neutron star—a pulsar, as introduced earlier in this chapter. Others showed no such pulses. Furthermore, astrophysicists can compute the masses of stars in a binary system, and in some systems the mass of the companion star exceeds theorists' upper limits for the mass of a neutron star. Here, black holes provide the simplest and most coherent explanation of the observations.

Evidence for black holes in binary systems grew stronger through the late twentieth century, and today most astrophysicists acknowledge the existence of black-hole binaries. Although these systems are too distant for us to see their details directly, decades of increasingly finer observations and theoretical modeling have given us a firm picture of what a black-hole binary must look like. As Figure 15.6 shows, the salient features are a large, visible star and an invisible black hole in close orbits around each other. Star and hole are so close that gas flows from the former to the latter. Because of the system's orbital motion, the gas doesn't flow straight but forms a swirling, donut-shaped cloud around the black hole. This *accretion disk* is where the intense heat and x-rays are generated. The gas loses energy through the heat-generating friction, allowing it to spiral ever closer to the black hole.

Fig. 15.6 What a black hole in a binary system might look like. The massive star in the center is distorted by the strong tidal forces of the black hole. Gas from the star is drawn to the vicinity of the hole, where it swirls around in a disk before finally disappearing into the hole itself. Friction in the gas generates such high temperatures that the gas produces copious x-ray emission.

Even as some astrophysicists studied the newly found binary x-ray sources in our own galaxy, others sought to understand some of the most puzzling and distant objects ever found—the quasars. These enigmas first appeared as pointlike images at the very edge of the visible Universe, yet they emit more energy than an entire galaxy comprising hundreds of billions of stars. What could possibly power such colossal energy sources?

Higher resolution images soon showed that quasars' energy generation regions must be very small, and again astrophysicists' imaginations turned to black holes. Might the accretion of matter into the hole account for a quasar's extreme energy output? Another clue came from images showing some quasars emitting high-speed jets of material. A newly discovered solution for the structure of spacetime around a rotating black hole suggested that such jets would be a natural feature of a rotating hole system, with matter accreting near the "equator" of the spinning hole and jets emerging at its "poles." Furthermore, quasars seemed similar to so-called active galaxies, which also exhibit jets and substantial energy output from their galactic centers. Could the same black-hole mechanism be at the heart of both sorts of objects?

With the advent of the Hubble Space Telescope and high-tech Earth-based telescopes late in the twentieth century, astronomers observing nearby galaxies found similar compact, high-energy sources at the centers of many galaxies, including our own. The high speeds of stars and gases swirling around these centers again suggest black holes as the ultimate energy source. By now, it appears that quasars, active galaxies, and normal galaxies like our Milky Way probably all contain massive central black holes. The quasars are believed to be the cores of galaxies so distant that we're seeing some as they were billions of years ago, at a time when accretion into their central black holes was particularly active and their energy output exceptionally strong. Active galaxies have less vigorous accretion rates, and in normal galaxies the process is quite modest.

How big are the black holes at galactic centers? The hole in our Milky Way contains several million Suns' worth of matter. Holes lurking in some quasars may exceed a billion solar masses. In physical dimension, these holes would be about 10 times the diameter of our Sun and roughly the size of the Solar System, respectively. Should you decide to explore these holes by dropping freely into them from a great distance, you would have about 10 seconds between crossing the horizon and reaching the singularity of the Milky Way's hole and a couple of hours for the quasar's hole. Of course these black holes continue to grow, on astronomical timescales, by gobbling up stars and gas from the high-density surroundings of their galactic-center neighborhoods.

Ripples in Spacetime

Drop a rock into a pond and circular ripples spread across the water. Only later, when the ripples have reached it, can a distant point on the pond "know" about the rock's plunge. Disturb spacetime, perhaps through a violent collision between black holes, and what happens? Special relativity assures us that distant points can't know instantaneously about this event; indeed, that's one of the reasons Einstein knew that Newton's theory of gravity couldn't be right. As on the pond, "ripples" in spacetime itself propagate outward, carrying information about the violent event at their center. What's a ripple in spacetime? Simply a change in the curvature of spacetime—a change that moves outward from the disturbance that initiates it. Einstein's general relativity predicts the existence of these ripples in spacetime; they're called *gravitational waves*. General relativity also shows that gravitational waves travel at the familiar speed of light, c.

Gravitational waves are an entirely new phenomenon predicted by general relativity. Detecting these waves would provide an independent confirmation of Einstein's theory, and might give us a novel window on the cosmos. How can we detect them? With their peaks and troughs, gravitational waves would stretch and compress spacetime itself. We could detect the spatial part of that stretch and compression by measuring the associated motion of physical objects. Because mass is what responds to gravity—that is, to spacetime curvature—we'll have better luck with massive objects. For some decades researchers around the world have attempted to detect gravitational waves by using huge aluminum bars equipped with exquisitely sensitive motion detectors. The bars would be set into vibration by a passing gravitational wave. To eliminate vibrations induced by trucks, scientists walking by, and other mundane causes, a typical experiment involves identical setups located thousands of miles apart. The only events considered real candidates for gravitational waves are those that trigger both detectors. Although these experiments have produced a few intriguing signals, none to date has passed muster as a true gravitational-wave detection.

This may all change soon, however, as a new generation of gravitational-wave detectors becomes operational. Abandoning the massive cylinders of first-generation detectors, gravitational-wave researchers are now turning to interferometry—the method pioneered by Michelson in his famous experiment with Morley—to measure precisely the distance between two widely separated objects. Recall from Chapter 6 that a slight change in the travel time for light along one arm of the Michelson–Morley apparatus (shown in Figure 6.2) would result in a shift in the observed interference pattern. Michelson and Morley hoped to find changes associated with differences in the speed of light; for gravitational-wave detection, we're looking for changes in the distance from beam splitter to mirror as a spacetime ripple goes by.

In the United States, the Laser Interferometer Gravitational-Wave Observatory (LIGO) consists of two complete interferometers, one in Washington state and one in Louisiana. Each consists of a Michelson-type apparatus with two perpendicular arms 2.5 miles long. These instruments can measure changes in the lengths of these 2.5-mile paths of less than a trillionth the diameter of a human hair, and with that sensitivity they should be able to detect gravitational waves produced in the supernova explosions that result in neutron stars and black holes; in collisions of black holes and neutron stars; and in the Big Bang explosion that began our Universe. Similar detectors are being built around the world, and collectively they will give astrophysicists a new type of "telescope" for observing hitherto unseen events in the cosmos.

Even more ambitious is the proposed Laser Interferometer Space Antenna (LISA), a Michelson-type apparatus whose "arms" will consist of spacecraft forming a triangle 3 million miles on a side (that's 400 times Earth's diameter, or 3 percent of the Earth–Sun distance!). Sensitive to changes in that distance on the order of a billionth of an inch, LISA should "see" gravitational waves from beyond our galaxy, including those generated by the massive black holes at the centers of other galaxies.

Although gravitational waves have yet to be detected directly, astrophysicists nevertheless have one piece of convincing evidence for their existence. This is the Taylor–Hulse binary pulsar, which I introduced early in the chapter for its obvious general-relativistic orbit precession. Recall that the orbital period of this binary neutron star admits very precise measurement and that this period is changing slowly with time. Why changing? Because the neutron stars' orbits are shrinking. Why shrinking? Because the stars are losing energy, gradually spiraling closer just as a satellite in orbit near Earth slowly loses altitude through friction with the upper atmosphere. But there's no atmospheric friction in the binary pulsar. Instead, the neutron stars lose energy through an unseen process that leaves its fingerprint in the slowly decaying orbital motion. That process is the generation of gravitational waves. As the massive neutron stars swing round in their close orbit, the energy they expend in disturbing spacetime—energy that's carried away as the yet-undetected ripples of gravitational waves—is lost from their orbital motion. Years of meticulous observation show that the binary pulsar is losing energy at just the rate that general relativity predicts. So although we haven't "seen" gravitational waves from the binary pulsar, we're quite sure they're being produced and are responsible for the observed orbital changes. Incidentally, their meticulous observations of the binary pulsar and its general relativistic implications earned Taylor and Hulse the 1993 Nobel Prize in Physics.

Relativity in the Astrophysicist's Toolbox

The observational technologies and physical theories behind modern astrophysics reveal a Universe rich in strange and wondrous phenomena, some so strange that they would have seemed unimaginable only a few decades ago. Many of these phenomena—from black holes to binary pulsars, from gravitational waves, lenses, and time dilation to unusual orbits in strongly warped spacetime—are solidly confirmed and regularly observed. Others, like the wormholes of Carl Sagan's *Contact*, remain speculative. They have some grounding in theory, though no supporting observational evidence. But all these phenomena share common roots in the extraordinary, flexible spacetime geometry unveiled by Einsten's relativity. For that reason, relativity has become a working tool for many of today's astrophysicists. Even those who don't observe or theorize about relativistic objects may still find relativistic tools useful, as when gravitational lenses aid in the search for Earthlike planets circling distant stars. Relativity remains the ultimate tool for those who seek to understand the really Big Picture—the workings of the entire Universe. We'll go there in the final chapter.

CHAPTER 16

EINSTEIN'S UNIVERSE

● ● ●

The Big Picture

Gravity isn't just about planetary orbits or even exotic black holes. Gravity binds stars into galaxies, galaxies into clusters of galaxies, and clusters into superclusters. Structure persists to the largest scales, with "walls" and "bridges" where galactic superclusters concentrate. Between these are sparsely populated voids. Had we evolved in a galaxy in such a void, it would not have been until the end of the twentieth century that our astronomical technology would have been capable of detecting even the nearest neighbor galaxies. But then, what incentive would we have had to develop that technology?

How did structure evolve in a Universe that, at its beginning, was a simple, homogeneous soup of matter and energy? What is the shape of the Universe today? What is its ultimate fate? On the vast scales of distance and mass that describe the Big Picture configuration of today's Universe, gravity is the dominant influence. Our description of gravity—the general theory of relativity—is therefore at the heart of cosmology, the study of the overall structure of the Universe.

Einstein's Blunder

Einstein himself was among the first to apply general relativity to cosmology, and what he found was unsettling. The Universe, according to the simplest formulation of general relativity, couldn't be static; it had to be expanding or contracting. But prevailing wisdom held that the Universe *was* static, having existed forever unchanged in its overall features. Now, in the mathematical development of general relativity, there arose a number called the *cosmological constant*, whose value seemed to be arbitrary. Absent any reason to the contrary, the most sensible choice is to set this number to zero, giving the simplest formulation of the theory. That choice is what made Einstein's universe expand or contract, so he introduced a nonzero cosmological constant of just the right value to keep the Universe static. Einstein's cosmological constant represented a sort of repulsive force acting on the largest scales, preventing the Universe from collapsing under its own gravity. Nothing then known about gravity suggested such a repulsion, so Einstein was a little uneasy with his cosmological constant. But it was not inconsistent with the general theory, so he accepted the cosmological constant as necessary to make his theory fit what seemed to be the real Universe.

Other physicists also explored relativity-based models for the Universe. In the early 1920s, the Russian Aleksandr Aleksandrovich Friedmann found solutions to the equations of general relativity that described an evolving universe beginning in a very dense state and then expanding at an ever slowing rate. Friedmann's results revealed two distinct kinds of possible universes: those in which expansion continues forever, albeit at an ever slower rate; and those in which the expansion eventually halts and the universe subsequently contracts to an eventual high-density crunch. These two possible fates are intimately linked to the overall spacetime geometry of the Universe. A forever expanding Friedmann universe has negative curvature, meaning its spacetime is shaped like the four-dimensional analog of a saddle or a pass between mountain peaks, and is infinite in extent (see Figure 16.1a). A universe that eventually collapses has positive curvature, like the surface of a sphere (Figure 16.1b). It's closed back on itself and is finite in extent. Yet, like a sphere's surface, it has no edge. Between these two types of Friedmann universe is a dividing case. Overall, it's flat (although spacetime in such a universe would still be curved locally in the vicinity of massive objects), infinite, and will just barely expand forever.

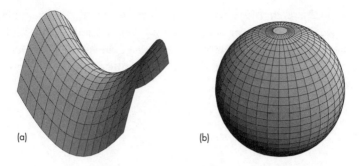

Fig. 16.1 (a) A universe with negative curvature would be analogous to this saddle-shaped surface in three dimensions. (b) A positively curved universe would be analogous to a three-dimensional sphere.

These geometrical analogies are imperfect. You can imagine a sphere or saddle shape by picturing it in three dimensions, and I'm asking you to think of the four-dimensional spacetime Universe as being analogous to the surface of the sphere or saddle. But that "surface" is all there is; there's no fifth dimension for it to be curved in. And the four-dimensional Universe may be expanding. Into what, you ask? No answer; it's the fabric of space and time itself that's expanding.

By the late 1920s, the American astronomer Edwin Hubble had completed a study of nearby galaxies and found that they were receding from us at speeds directly proportional to their distances. Hubble concluded that our Universe is indeed expanding. (If this sounds like we're at the center—a disturbingly un-Copernican thought—rest assured. In an infinite Universe there's no center, and if ours is a finite Friedmann universe there's still no center—just as there's no point on the *surface* of the Earth that can claim to be the center.) So the cosmological constant wasn't necessary after all, and in the 1930s Einstein abandoned the constant as "the greatest blunder of [his] life."

Through the rest of the twentieth century, astrophysicists accumulated ever more evidence, both observational and theoretical, for a Universe that began in a Big Bang explosion some 14 billion years ago and has been expanding ever since. Much of this evidence lies outside the realm of relativity, and I'll leave it to others (see the Further Readings list) to present that story. Yet, at the heart of modern cosmology lies general relativity, because it's the interaction between matter and spacetime that shapes the overall geometry of the Universe.

The Fate of the Universe

One big question in cosmology is whether the Universe will expand forever or will ultimately contract. Absent a cosmological constant, general relativity shows that this question is equivalent to another: What is the large-scale curvature of spacetime? If it's positive, like the surface of a sphere, then the Universe is closed and will ultimately contract to a Big Crunch. If it's negative, like a saddle-shaped surface, then the Universe is open and will expand forever. And if it's flat—meaning the rules of tenth-grade Euclidean geometry apply on the large scale—then the Universe will just barely expand forever.

What determines the geometry and thus the ultimate fate of the Universe? Because matter is what warps spacetime, giving it curvature, the answer lies in the amount of matter present in the Universe. There's a *critical density*, averaging on the order of a few hydrogen atoms in each cubic meter of otherwise empty space, above which the amount of matter is enough to curve spacetime back on itself, giving a closed, finite, positively curved Universe that will eventually collapse. If its density is below critical, the Universe will be open, negatively curved, and will expand forever. Right at the critical density, the overall geometry of the Universe is flat and it will just barely expand forever.

So which is it? Remarkably, the real Universe seems close enough to the dividing case of critical density that the answer isn't obvious from observations. The visible matter—stars, interstellar gas clouds, galactic centers, and everything else that emits electromagnetic radiation—constitutes only a small fraction of the matter needed to achieve critical density. However, the movement of stars within galaxies suggests that unseen *dark matter* also contributes to the Universe's overall density. What is that dark matter? Some of it could be mundane things like burned-out stars too faint to see, or isolated black holes, but theory suggests that most of it must be in the form of undiscovered exotic matter different from the ordinary matter that we see throughout the Universe. Here's a humbling thought: perhaps most of the Universe is made of something we know essentially nothing about!

Today, most cosmologists believe there's enough dark matter that the Universe has exactly the critical density and that its overall geometry is flat. A number of theoretical arguments suggest a flat Universe, and recent studies of the so-called cosmic microwave background radiation—a kind of fossil remnant of times early in the Big Bang—support the flat-Universe hypothesis.

That would end the Big Picture story of spacetime geometry, except for a remarkable discovery made in 1998. Astrophysicists had been studying supernova explosions in very distant galaxies, hoping to use their measured brightness to infer the rate at which the expansion of the Universe is slowing. That method works because distant objects appear from Earth as they were a long time ago, as long ago as it took light from them to reach us. Comparing the brightness and recession speed of supernovas at different distances thus gives the expansion rate of the Universe at different times. To the astrophysicists' great surprise, they found that the expansion wasn't slowing at all; rather, it was accelerating! It appears that this acceleration began about 5 billion years ago, coincidentally around the time the Sun, Earth, and the rest of our Solar System formed.

What could cause an acceleration of the cosmic expansion, despite the slowing effect of mutual gravitation? One possible answer goes right back to Einstein: His cosmological constant provides a cosmic repulsion that could increase the expansion rate. That's only one possible answer, and cosmologists are proposing new phenomena with names like "quintessence" and "dark energy" that could also produce an accelerated cosmic expansion. With more detailed observations of ever more distant realms becoming accessible to our telescopes, and after much theoretical analysis, we'll eventually learn whether Einstein was right after all when he introduced the cosmological constant into general relativity.

A Theory of Everything?

General relativity provides our best understanding of the Big Picture of the Universe as a whole, but can general relativity tell us everything? The answer is no. That's because physicists have not yet been able to reconcile general relativity with quantum physics, the theory that describes matter on atomic and smaller scales. *Special* relativity and quantum physics were reconciled decades ago, giving us a powerfully accurate description of the behavior of matter at small scales that reveals some entirely new atomic phenomena required by the Principle of Relativity. A reconciliation of general relativity and quantum physics, though, faces deep conceptual problems. That's because the essence of quantum physics is *quantization,* meaning that the "stuff" of the Universe—from particles of matter to energy itself—comes in discrete "chunks" rather than being continuously subdividable. You can have one electron, but not half of one. You can have a "chunk" of light energy of a given color (called a *photon*), but you can't have less. It's this essential graininess that ultimately dictates the strange rules governing the quantum world.

We, and the things we interact with in everyday life, are so large that we don't notice quantization. A glass of water contains so many individual water molecules that the quantization of the water into molecules doesn't seem to make a difference; the water might just as well be a continuous fluid. A light bulb or even a candle flame emits so many photons that they might as well constitute a continuous stream of energy. All this is even more true for the planets, stars, and galaxies that make up the astrophysical Universe. As a practical matter, astrophysics' description of the Universe is at such a large scale that the reconciliation of quantum physics and general relativity is usually unimportant. Put another way, the curvature of spacetime is generally significant only on scales vastly larger than the size of atoms or elementary particles. But we can imagine situations where this is not true, situations where spacetime is so tightly curved that even something as small as an elementary particle is big enough to experience spacetime curvature. What absurd situations would those be? One is the singularity at the center of a black hole. There, general relativity predicts that spacetime curvature becomes infinitely sharp. Before that true singularity is reached, effects of quantum physics must come into play. Another example is the very early Universe, at the start of the Big Bang—specifically, the time before about 10^{-43} of a second from the beginning. (That's 1/10 000 000 000 000 000 000 000 000 000 000 000 000 000 000 of a second, a time known as the *Planck time*.) At that point the Universe was so dense that even its geometrical structure would have to be described using quantum physics. Because we don't know how to reconcile quantum physics and general relativity, we can't say much about conditions at the center of a black hole or in the early Universe before the Planck time. Ultimately, our knowledge is incomplete.

Why are the two pillars of modern physics—general relativity and quantum physics—so seemingly irreconcilable? Because general relativity is, at its heart, a continuous theory. It envisions a spacetime that is continuously divisible. That means we can divide a meter or an inch into ever smaller lengths, without limit. Similarly, we can divide the 1-second interval between ticks of a clock into as many ever tinier sub-intervals as we wish. For time intervals much longer than the incredibly tiny Planck time introduced in the previous paragraph and for distances much longer than the *Planck length*—or the distance that light travels in one Planck time (about 10^{-35} of a meter)—the idea of a continuous spacetime is a very good approximation to reality. As we approach the Planck length and time, though, spacetime itself must exhibit quantization. There must be fundamental, indivisible units of length and time on the order of the Planck units. What does this mean? No one knows for sure. Because the quantum world is seething with submicroscopic events, physicists often picture quantized spacetime as a spongelike structure with a complicated and everchanging geometry, in which wormholes and bridges continually form and dissolve.

Will we ever reconcile general relativity and quantum physics, producing a successful theory of *quantum gravity*? Probably. Some physicists believe we're close, with a group of theories collectively called *string theory* or *M-theory*, which envision the fundamental entities of nature not as particles but as tiny looplike strings. Vibrations of these strings represent the different elementary particles of physics. Because the strings have finite size, even though they're indivisible, they manage to sidestep problems with gravity at the Planck scales. Furthermore, some versions of the theory suggest particles called *gravitons* that would quantize gravity in the same way the photon is a quantized chunk of electromagnetic energy. No string theory yet conforms to the observed set of elementary particles and all require a remarkable complication—the inclusion of at least six more dimensions in addition to the familiar three dimensions of space and one of time. Furthermore, no one has yet figured out how to test experimentally most predictions of string theories. For some physicists, string research is nevertheless getting us close to a "Theory of Everything," explaining the behavior of the entire Universe from the smallest to the largest scales. To others, string research seems a fruitless mathematical exercise. Time will tell.

Out of the Black Hole: Inflation, Multiple Dimensions, and Parallel Universes

Although no one has yet figured out how to merge general relativity and quantum physics into a consistent theory, we do have some hints of how the quantum realm might influence relativity. The British physicist Stephen Hawking—occupant of Newton's former chair at Cambridge University—has shown how quantum effects lead to a subtle "glow" of radiation from the spacetime around a black hole. That radiation ultimately saps the black hole of its energy—equivalently, by $E = mc^2$, of its mass—causing the hole to "evaporate" and eventually disappear altogether. For the massive black holes that astrophysicists have discovered, that process will take far longer than the present age of the Universe. However, tiny black holes that might have formed early in the Big Bang would already have vanished through this mechanism of *Hawking radiation.*

Quantum physics shows that the emptiness we call vacuum isn't quite empty, but is seething with ghostly particle–antiparticle pairs that burst into a fleeting existence and promptly annihilate. Hawking radiation arises when one of these particles falls into a nearby black hole, leaving its partner no one to annihilate with. The lonely partner then looks like a particle that's emerged from the hole. Although that isn't quite the case, the black hole has, in a sense, created the new particle and thus has given up some of its own mass-energy.

Not only is the seemingly empty vacuum actually the site of frenetic physical activity, but the current theory of elementary particle physics—the so-called standard model—also suggests the possibility of a "false vacuum," a high-energy state that might decay rapidly to a true vacuum. Decay of the false vacuum can lead to exponentially rapid inflation of spacetime. Detailed theories of the Big Bang suggest that a period of just such inflation occurred in the first fraction of a second of the Universe. During this brief instant, the Universe grew in size by many orders of magnitude—a process that flattened out any curvature and resulted in a Universe whose large-scale spacetime geometry is essentially the flat, Euclidean geometry that you learned in tenth grade.

It's the prospect of inflation that leads to a remarkable possibility I mentioned in Chapter 1: a baby universe could bud from a parent and then grow by inflation to become a full-blown universe in its own right. Repeated endlessly through eternity, this process would give rise to a complex Multiverse of which our Universe is but one branch.

Attempts to merge relativity with quantum physics often seem to lead us beyond the three familiar dimensions of space and one of time. Today's eleven-dimensional string theories are only the latest instance of multidimensional theories of everything. Einstein himself spent much of his later life working, in vain, on five-dimensional theories that he hoped would combine his general relativity with quantum physics.

Extra dimensions allow for the remarkable possibility that there may be parallel universes lying very close to ours but separated by an extra dimension that we don't directly perceive. Stanford physicist Savas Dimoupolos and colleagues have argued that a parallel universe might be only a few millimeters (about a tenth of an inch) away from ours (Figure 16.2). We think we know nothing of it because all processes in that universe are confined to its three dimensions of space and one of time. But not quite! According to Dimoupolos, the quantized particles of gravitational

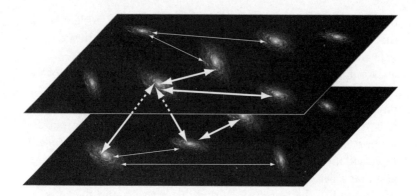

Fig. 16.2 Parallel universes? Each plane is a two-dimensional analog of a universe with three space dimensions, populated with galaxies. The planes shown are millions of light-years in extent, but they're separated by a fraction of an inch in an extra dimension that we don't directly perceive. Thin arrows represent electromagnetic interactions (e.g., light beams) that are confined to the individual universes. Gravity (thick arrows) extends into the extra dimension, so each universe feels the other's gravitational effect. (Galaxy photos courtesy of NASA.)

influence—the gravitons—would be unique in that they could traverse the extra dimension. So, while ordinary matter, light, and other forms of energy would be confined to that nearby but hidden universe, its gravitational influence, as Figure 16.2 shows, would not. If this idea is correct, then what we infer to be dark matter in our Universe may actually be ordinary matter in the parallel universe! We perceive only its gravitational influence, not the matter itself, and so of course it seems dark to us. Strange as this parallel-universe idea is, no one has yet disproved it, and experiments with gravity over millimeter scales might just reveal that extra dimension.

Because spacetime can be curved, it's also possible that the parallel universe isn't really a distinct universe but rather a part of ours that's very far away in the ordinary dimensions but close in the extra dimension. If that's true, then it's possible to imagine a short wormhole connecting our neighborhood to the distant realm of the parallel universe. Maybe that's what the machine in Carl Sagan's *Contact* is all about.

Intuiting Relativity

We've now come full circle, back to some of the stranger ideas I introduced briefly in the first chapter. Those ideas, and other new concepts ranging from the relativity of simultaneity to black holes and gravitational waves, become possible when we abandon rigid, absolute space and time and replace them with the flexible, curving spacetime of Einstein's relativity. Einstein's special and general theories of relativity reveal a Universe far richer and stranger than anything in our commonsense experience. In this book I've taken you rather thoroughly through the special theory, trying to convince you of its validity on rigorous logical grounds. Yet I've continually reminded you that you'll never have a fully natural intuition for relativity. I don't either, and I don't believe Einstein did. That's because none of us has experienced in everyday life the conditions that show up the difference between Einsteinian relativity and Newtonian common sense.

With general relativity I've had to leave things a lot murkier. I've given you the briefest motivation for the fundamental underpinnings of the theory, ultimately leading to the idea of gravity as the geometry of spacetime. In these final chapters I've described some of the consequences of general relativity—especially significant where gravity is strong—but I haven't at this level been able to give a lot of motivation for those consequences, let alone any kind of intuitive feel for them.

So you can understand relativity logically, and, if you work with it enough, gain a comfortable familiarity. For the nonscientist that's possible with the simpler concepts and mathematics of special relativity, and for the mathematically brave a thorough understanding of general relativity is possible too. But in neither case will you have much natural, commonsense intuition about relativity. Is such intuition at all possible? I believe it would be if we grew up experiencing the Universe in the full richness that relativity describes. If, as a baby, you crawled at speeds approaching that of light, then your common sense would be fully consistent with special relativity. You would have no misleading notions about absoluteness of space and time measurements, and reference frame–dependent lengths, times, and even simultaneity would be the norm for you. You wouldn't be surprised when a friend went out for a high-speed jog and came back a member of a younger generation. In short, special relativity would make perfect sense to you because it would be part of your regular experience.

If, in addition to being a relativistic crawling baby, you were so large that your body directly experienced the curvature of space-time, then the geometrical nature of gravity would also be intuitively obvious to you. Your geometry teacher, drawing on a blackboard so big that spacetime curvature was obvious, would never deceive you with Euclidean nonsense; it would be obvious that the angles of a triangle add to other than 180 degrees and that parallel lines might meet (or might not, depending on the sign of the local spacetime curvature). You would experience directly that these geometrical effects arose in the presence of matter and energy, and your understanding of gravity would naturally be that of Einstein. But you're small in relation to the spacetime curvature in your neighborhood and you move slowly relative to things important in your life; thus, for you, relativity can never seem intuitive. You can, however, grasp the simple principle at the heart of all relativity—that motion doesn't matter, or that there's no preferred frame of reference—and despite the failure of your intuition, you can understand intellectually the remarkable consequences of this principle and appreciate the wonderfully rich Universe it engenders.

TIME DILATION

● ● ●

Here I'm going to use high-school math to convince you that the formulas for time dilation and length contraction do indeed follow directly from the Principle of Relativity. You can take those formulas on faith and skip this appendix, but if you're a stickler for logical consistency, then reading this will give you what you need.

High-School Math Review

At some point in your school career, either in junior high or early high school, you learned the *Pythagorean theorem*, which gives the length of the hypotenuse (long side) of a right triangle in terms of the lengths of the other two sides. Figure A.1 shows a right triangle with sides labeled A, B, and hypotenuse C. The Pythagorean theorem says

$$C^2 = A^2 + B^2.$$

You also learned that you can do the same thing to both sides of an equation and it's still a valid equation. So let's take the square root of both sides, to get the length of the hypotenuse, C:

$$C = \sqrt{A^2 + B^2}.$$

I've marked this value on Figure A.1.

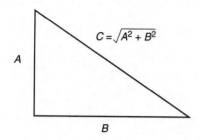

Fig. A.1 The Pythagorean theorem gives the hypotenuse of a right triangle in terms of the lengths of the other two sides.

You also learned, probably even before high school, that *distance = speed × time*. After all, that's just what speed means: go 50 miles per hour for 2 hours, and you've gone 100 miles. Distance equals speed times time.

Time Dilation

Now you're all equipped to understand time dilation, quantitatively. Figure A.2 is a modified version of Figure 8.3, the light box that I used to introduce time dilation. I've emphasized the light path by making it a solid line and dimming everything else. In Figure A.2a

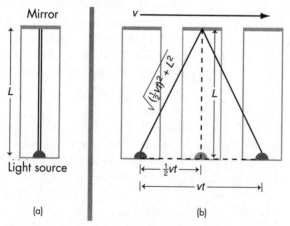

Fig. A.2 The light box of Figure 8.3, shown with various quantities used to derive the time-dilation formula.

the box is again shown in a reference frame where it's at rest, and it's clear that the light goes a round-trip distance $2L$ between source, bouncing off the mirror, and returning to the source. We've called the time in this reference frame t' ("t prime"), and the light's speed is, of course, the speed of light, c. So the formula *distance = speed × time* becomes

$$2L = ct'.$$

(Here I'm adopting the convention of implied multiplication, where two symbols written next to each other are assumed to be multiplied. This avoids use of the multiplication sign.) Dividing both sides of this equation by c gives

$$t' = \frac{2L}{c}.$$

That was easy!

Now let's go to the frame in which the box is moving (Figure A.2b). The box is moving at speed v, and we're calling t the time between departure and return of the light flash. So *distance = speed × time* tells us that the distance the box travels is just vt. While the box goes *half* this distance, the light itself takes the diagonal path up to the mirror, as shown on Figure A.2b. How long is this diagonal? It's the hypotenuse of a right triangle with sides $\frac{1}{2}vt$ and L. The Pythagorean theorem then gives its length:

$$\sqrt{(\tfrac{1}{2}vt)^2 + L^2}, \quad \text{or} \quad \sqrt{\tfrac{1}{4}v^2t^2 + L^2}.$$

The total light path is *twice* this distance, the time the light takes is t, and—here's where relativity comes in—the speed of the light is c in this reference frame as well. So *distance = speed × time* now gives

$$2\sqrt{\tfrac{1}{4}v^2t^2 + L^2} = ct.$$

Now, we want to know the time t. Unfortunately it's on both sides of this equation and tangled up in a square root. To get it out, square both sides:

$$4(\tfrac{1}{4}v^2t^2 + L^2) = c^2t^2, \quad \text{or } v^2t^2 + 4L^2 = c^2t^2.$$

Now subtract v^2t^2 from both sides:

$$4L^2 = c^2t^2 - v^2t^2, \quad \text{or } 4L^2 = c^2t^2\,(1 - v^2/c^2).$$

Almost done! Divide both sides by c^2 and take the square root of both sides. The result is

$$\frac{2L}{c} = t\sqrt{1 - v^2/c^2}.$$

But $2L/c$ is just the time, t', measured in the frame at rest with respect to the box [see Equation (A2)]. So we have

$$t' = t\sqrt{1 - v^2/c^2}.$$

This is just the time-dilation formula I presented in Chapter 8, except that it has c in it. In Chapter 8 I defined the speed v as a fraction of c, so the quantity v/c here is the v of Chapter 8. Thus we've reached Chapter 8's formula, $t' = t\sqrt{1 - v^2}$, and we've done so in a way that follows rigorously from the Principle of Relativity.

GLOSSARY

• • •

Aberration of starlight The change in apparent position of a star, due to Earth's orbital motion. Used to show that Earth did not drag ether with it.

Acceleration The rate at which an object's motion changes. Acceleration includes changes in speed or direction.

Accretion disk The disk-shaped cloud of matter swirling around and into a black hole.

Action at a distance The Newtonian view that influences, especially gravity, reach instantaneously from a gravitating object to more distant objects. The action-at-a-distance picture is inconsistent with special relativity.

Antimatter Matter consisting of particles with properties that are exactly the opposite of ordinary matter. Antielectrons (positrons), for example, are like electrons but with positive charge. Antimatter is created in energetic interactions involving elementary particles or high-energy electromagnetic radiation. When they meet, matter and antimatter annihilate in a burst of energy.

Beam splitter A device that splits a beam of light into two beams traveling on different paths. The simplest beam splitter is a mirror with insufficient reflective material.

Big Bang The cosmic explosion that began the Universe.

Big Crunch In theories that propose an oscillating Universe that expands and then contracts, the Big Crunch is the ultimate state of contraction before the next expansion. In the Big Crunch, all matter would be compressed to near-infinite density.

Black hole An object so massive yet so compact that not even light can escape its gravity.

Cosmological constant A number Einstein introduced into the general theory of relativity so the theory would predict a static Universe. Einstein abandoned the cosmological constant when observations showed that the Universe was in fact expanding. Discoveries at the end of the twentieth century suggest the constant might be necessary after all.

Cosmology Study of the origin, evolution, and large-scale structure of the Universe.

Critical density The minimum average density necessary for the Universe to expand forever instead of eventually collapsing.

Dark matter Unseen material believed to make up over 90 percent of the Universe's mass.

Doppler effect The increase (or decrease) in the frequency of waves—sound or light—when the source of the waves is moving toward (or away from) the observer.

Electric charge A fundamental property of matter that is at the basis of all electric and magnetic interactions.

Electromagnetic induction The phenomenon whereby a changing magnetic field produces an electric field.

Electromagnetic wave An electromagnetic phenomenon in which changing electric and magnetic fields continually generate each other, producing a wave of electromagnetism that travels with the speed of light, c. Electromagnetic waves include radio waves, microwaves, infrared, visible light, ultraviolet, x-rays, and gamma rays.

Electron A fundamental particle in nature. Carries a negative electric charge and little mass.

Elsewhere The elsewhere of an event consists of those events that cannot influence or be influenced by the given event.

Energy One of the two manifestations of mass-energy, the fundamental "stuff" of the Universe. Energy takes many forms, including the energy of motion, gravitational energy, heat energy, electromagnetic energy, etc. Without energy there would be no motion, no activity whatsoever.

Epicycle Smaller circular paths of the planets, imposed on their larger circular orbits and required in early cosmological models to account for retrograde motion.

Escape speed The speed necessary for an object to escape to an infinitely great distance from a gravitating body. At Earth's surface, escape speed is about 7 miles per second; at the horizon of a black hole, it's the speed of light.

Ether Hypothetical substance proposed by nineteenth-century physicists as the medium of which light waves are a disturbance.

Event Something that happens, characterized by where it occurs in space and when it occurs in time.

Event horizon The region around a black hole at which escape speed becomes the speed of light. Nothing—no material object, no light, and no information—can escape from within the event horizon.

Field An invisible influence in the space surrounding a massive object (gravitational field), an electrically charged object (electric field), or a moving charged object (magnetic field).

Force A push or pull, either by an obvious agent or an invisible influence like gravity, electricity, or magnetism. In Newtonian physics, forces cause changes in motion.

Frame of reference The surroundings that share one's state of motion; the physical setting that establishes one's point of view for making measurements.

Free fall, free float Terms applied to a situation in which an object moves under the influence of gravity alone.

Frequency The number of times per second that a periodic occurrence, such as the oscillation of a wave, takes place.

Future In relativity, the future of an event consists of all those events that the given event can influence.

Galilean relativity The principle, known since the time of Galileo and Newton, that the laws of motion provide no way to distinguish different states of uniform motion. In other words, the question, Am I moving? is meaningless as far as the laws of motion are concerned.

General relativity Einstein's 1916 theory that describes gravity as the curvature of spacetime.

Geodesic The straightest possible path through a space or spacetime that may itself have curvature. Great circles are the geodesics on Earth's surface.

Gravitational lens Any massive body whose gravitation, or spacetime curvature in general relativity, bends light from more distant objects. The result may be brightened, distorted, or multiple images.

Gravitational time dilation The phenomenon whereby time passes more slowly near a gravitating object. Also called gravitational redshift.

Gravitational wave A wave disturbance in the structure of spacetime itself. Gravitational waves originate in the acceleration of massive objects and propagate at the speed of light.

Hawking radiation A phenomenon associated with quantum physics in the presence of a black hole, whereby particle–antiparticle pairs come spontaneously into existence just outside the hole. When one of the pair falls into the hole, the other is left alone and appears as if it were radiation emerging from the hole. Eventually, Hawking radiation saps black holes of their mass-energy.

Interference A wave phenomenon whereby two waves meeting at the same place simply add. In constructive interference, the wave disturbances are in the same direction (e.g., crests meet crests, troughs meet troughs) and the effect is a strengthened wave. In destructive interference, crests meet troughs and the overall wave is diminished.

Inertial reference frame Any reference frame in which the law of inertia is obeyed. In practice, the only inertial frames are those in free fall (or free float).

Length contraction The reduction in the length of an object as measured by an observer with respect to whom the object is moving.

Light A form of electromagnetic radiation visible to the human eye. More loosely, in this book, the term is often used to designate any electromagnetic radiation.

Light-year The distance light travels in 1 year.

Luxon A particle that travels at exactly the speed of light relative to any uniformly moving reference frame. The photon, or quantized bundle of electromagnetic wave energy, is one familiar example.

Maxwell's equations The four equations developed in the 1860s by James Clerk Maxwell that describe all phenomena of classical electricity and magnetism.

Mechanics The study of motion, one of the major branches of physics.

Medium The substance of which a given wave is a disturbance. Air, for example, is the medium for sound waves, and water for water waves. Light waves do not require a medium.

Michelson interferometer A device that uses interference of light waves traveling on two perpendicular paths to detect minute changes in speed or distance between the two paths.

Michelson–Morley experiment The famous 1887 experiment in which Michelson and Morley failed in their attempt to detect Earth's motion through the ether, despite using a Michelson interferometer more than adequate to the task. The failure of this experiment paved the way for Einstein's relativity.

Multiverse A multibranched system of multiple universes, proposed by some cosmologists as representing the overall structure of all that exists.

Muons Subatomic particles, created by cosmic rays high in the atmosphere and heading Earthward at nearly the speed of light, whose radioactive decay confirms time dilation.

Neutron star An object resulting from the collapse of a massive star and composed almost entirely of neutrons, with mass about that of the Sun's crammed into a sphere only a few miles across.

Newton's laws of motion Three laws describing the relation between motion and force. The first states that an object not subject to any force continues in uniform motion. The second states that an object's acceleration is proportional to the force applied to it and that for a given force the acceleration is less for a greater mass. The third law states that forces come in pairs; if one object exerts a force on another, then the second exerts a force of equal strength, but opposite direction, back on the first object.

Orbit The path described by an object moving under the influence of gravity alone.

Past In relativity, the past of an event consists of all those events that can influence the given event.

Photon A particlelike bundle of energy that, in quantum physics, is associated with electromagnetic waves.

Planck length The tiny length—about 10^{-35} meter—at which quantum physics should affect the nature of space itself.

Planck time The tiny time—about 10^{-43} second—at which quantum physics should affect the nature of time itself. To study happenings at the Planck time or length scale will require a successful merging of quantum physics with general relativity.

Positron Antiparticle to the electron.

Precession The gradual shift in the orientation of a planet's orbit. General relativity predicts that precession should occur but Newtonian gravitational theory does not. Precession has been detected in Mercury's orbit and in binary pulsars.

Present In relativity there is no such thing as a universal present, which would consist of events near and far that are simultaneous with what's happening here and now. Instead, the present of a given event consists, strictly speaking, of that event alone.

Principle of Equivalence The statement that it is impossible to distinguish the effect of gravity from acceleration and the absence of gravity from free fall.

Principle of Relativity The statement that the laws of physics do not depend on one's frame of reference. Thus absolute motion is a meaningless concept; only relative motion is meaningful.

Proper length The length of an object as measured in a frame of reference in which the object is at rest.

Pulsar A rapidly spinning neutron star that emits a searchlightlike beam of electromagnetic radiation, observed as a sequence of pulses as the beam sweeps repeatedly by Earth.

Quantum gravity A theory—as yet undeveloped—that would successfully combine quantum physics with general relativity.

Quasar A very distant astrophysical object emitting colossal amounts of energy. Quasars are believed to be the central regions of distant and therefore very early galaxies that emit copious radiation as material swirls into a massive black hole.

Retrograde motion The occasional reversal of direction that planets undergo when their motion through the sky is viewed from Earth.

Simultaneous events Events that occur at the same time. In relativity, events that are simultaneous in one frame of reference need not be simultaneous in other frames that are moving relative to the first.

Spacetime The union of space and time into a single four-dimensional structure.

Spacetime interval The four-dimensional "distance" between events in spacetime. The interval has the same value regardless of one's frame of reference.

Special relativity Einstein's 1905 theory based on the principle that the laws of physics are the same in all reference frames in uniform motion. The restriction to uniform motion is what makes this the *special* theory.

Standard model The theory that describes the different elementary particles and their interactions.

String theory A contemporary theory that envisions fundamental entities not as particles but as looplike strings. What we consider elementary particles would be different modes of vibration of these strings. String theory may have the potential for unifying quantum physics with general relativity, providing a theory of everything.

Tachyon A hypothetical particle capable of traveling faster than the speed of light, but not at light speed or slower. Tachyons have not been detected and their existence might wreak havoc on traditional notions of cause and effect.

Tardyon Any particle that travels at less than the speed of light relative to any uniformly moving reference frame. All known matter is composed of tardyons.

Time dilation A relativistic effect wherein the time between two events is shorter on a clock present at both events than it is when measured by two separate clocks in a reference frame where the events take place at different positions. Time dilation is sometimes described with the phrase "moving clocks run slow," but this is relativistically incorrect language for reasons described in the text.

Twins paradox A phenomenon resulting because time dilation occurs on both the outbound and return legs of a round-trip journey. A twin who makes such a journey returns to her starting point younger than her stay-at-home twin.

Uniform motion Motion in a straight line at constant speed.

Universal gravitation The idea, first developed by Newton, that every object in the Universe attracts every other object with a gravitational force that depends on the objects' masses and on the distance between them.

Wave A traveling disturbance that transports energy but not matter.

Wavelength The distance between adjacent wave crests.

Worldline The continuous path of an object in spacetime, tracing out the entire history of the object.

Wormhole A hypothetical "bridge" that would constitute a shortcut between two otherwise distant regions of spacetime, or even to another universe.

FURTHER READINGS

• • •

Brian, Denis. *Einstein: A Life*. New York: John Wiley & Sons, 1996.
A modern and thoroughly candid account of Einstein's life, among the first to exploit the Einstein Archives and Einstein documents made widely available in the 1980s.

Gott, J. Richard. *Time Travel in Einstein's Universe: The Physical Possibilities of Travel through Time*. Boston: Houghton Mifflin, 2001.
Gott, a Princeton physicist, not only elaborates relativity's clear implication that time travel to the future is possible but also explores the more speculative prospect of time travel to the past—a prospect that some physicists find increasingly worthy of serious study.

Greene, Brian. *The Elegant Universe: Superstrings, Hidden Dimensions, and the Quest for the Ultimate Theory*. New York: W. W. Norton, 1999.
Columbia University theorist Brian Greene gives a thorough and thoroughly readable account of string theory, the leading candidate for a "theory of everything" that would unite relativity and quantum physics. Greene's own contributions to the field are substantial, and he writes with an insider's deep knowledge and yet with the skill of one who can communicate complex ideas to nonspecialists.

Guth, Alan H. *The Inflationary Universe: The Quest for a New Theory of Cosmic Origins*. Reading, MA: Addison Wesley, 1997.
Guth is one of the originators of the inflationary universe idea, which resolves many of the conundrums raised by the Big Bang theory. This lively, accessible, yet authoritative account of our modern understanding of the origin of the Universe mixes solid science with the history of and personalities at the forefront of cosmology.

Hafele, J. C., and R. E. Keating. "Around the World Atomic Clocks: Predicted Relativistic Time Gains" and "Around the World Atomic Clocks: Observed Relativistic Time Gains." *Science* 177 (July 1972): 166–70.
This account of a real "twins" experiment done with highly accurate atomic clocks leaves no doubt about the reality of time dilation.

Livingston, Dorothy Michelson. *The Master of Light*. New York: Charles Scribner's Sons, 1973.

This biography by Michelson's daughter provides an insightful account of the Michelson–Morley experiment and its history, but also spares no detail of Michelson's often turbulent personal and professional life.

Pais, Abraham. *Einstein Lived Here*. New York: Oxford University Press, 1994.

A thematic, nonmathematical introduction to Einstein's life and thought. Author Abraham Pais, a physicist and Einstein scholar, has written extensively about Einstein's science and life. His earlier book, *Subtle Is the Lord: The Science and the Life of Albert Einstein* (Oxford University Press, 1982), delves deeply and quantitatively into Einstein's scientific work.

Paterniti, Michael. *Driving Mr. Albert*. New York: Dell Publishing, 2000.

Ever wonder what happened to Einstein after his death? His brain, at least, goes on—sloshing about in a Tupperware container! This amusingly improbable travelogue chronicles the true story of a cross-country drive with Einstein's brain and its eccentric steward.

Pyenson, Lewis. *The Young Einstein*. Boston: Adam Hilger Ltd., 1985.

A scholarly account of Einstein from childhood through the development of both relativity theories; particularly strong on the cultural and scientific milieu in which Einstein arose and on scientists' and mathematicians' reactions to Einstein's work.

Smolin, Lee. *Three Roads to Quantum Gravity*. New York: Basic Books, 2001.

A self-confessed optimist, physicist Lee Smolin believes we are only years away from unifying general relativity and quantum physics to produce a "theory of everything." In this lively and very contemporary book, he outlines three approaches that show promise.

Taylor, Edwin F., and John Archibald Wheeler. *Spacetime Physics*, 2d ed. New York: W. H. Freeman, 1992.

Written for sophomore-level college physics students, this book presents special relativity in a lively and entertaining style that emphasizes the underlying simplicity and fundamental principles of the theory. Some sections of the book are reasonably math-free and provide deep insights into relativity. If you don't mind a little algebra, *Spacetime Physics* will greatly enhance your understanding of Einstein's theory.

Thorne, Kip. *Black Holes and Time Warps: Einstein's Outrageous Legacy*. New York: W. W. Norton, 1994.

This definitive history of black holes and related phenomena of general relativity will convince you that black holes really do exist. Written in the first person by a leading relativity researcher, the book will reward a persevering reader with a really thorough but nonquantitative understanding of black holes in particular and relativity in general.

Will, Clifford. *Was Einstein Right? Putting General Relativity to the Test*. Basic Books, A Division of HarperCollins, 1986.

This highly readable book gives a thorough look at both the classical and fairly contemporary tests of general relativity. Written for the nonscientist, it's a fascinating blend of physics, personalities, and history.

Wambsganss, Joachim. "Gravity's Kaleidoscope." *Scientific American* 285 (November 2001): 64–71.
A nice account of gravitational lenses that shows how they work and the multiple uses to which astrophysicists are now putting them. Great illustrations, too!

INDEX

• • •